DEVELOPMENTS IN
HYDRAULIC ENGINEERING—5

CONTENTS OF VOLUMES 2, 3 AND 4

DEVELOPMENTS IN
HYDRAULIC ENGINEERING—5

Edited by

P. NOVAK

Emeritus Professor of Civil and Hydraulic Engineering,
University of Newcastle upon Tyne, UK

ELSEVIER APPLIED SCIENCE
LONDON and NEW YORK

ELSEVIER APPLIED SCIENCE PUBLISHERS LTD
Crown House, Linton Road, Barking, Essex IG11 8JU, England

Sole Distributor in the USA and Canada
ELSEVIER SCIENCE PUBLISHING CO., INC.
52 Vanderbilt Avenue, New York, NY 10017, USA

WITH 11 TABLES AND 89 ILLUSTRATIONS

© ELSEVIER APPLIED SCIENCE PUBLISHERS LTD 1988

British Library Cataloguing in Publication Data

Developments in Hydraulic Engineering.—5
—(Developments series).
1. Hydraulic engineering
I. Series
627   TC145

The Library of Congress has cataloged this serial publication as follows:
Developments in hydraulic engineering.—1.——London, England:
Applied Science Publishers, 1983–
v.: ill.; 22 cm.—(Developments series)
Published: London; New York: Elsevier Applied Science Publishers,

1. Hydraulic engineering—Periodicals.   I. Series.
TC1.D47   626′.05—dc19   86-657602

ISBN 1-85166-157-3

Typeset at the Alden Press, London, Northampton, Oxford

Printed in Great Britain at the University Press, Cambridge.

PREFACE

This is the fifth volume of *Developments in Hydraulic Engineering*, the publication of which started in the autumn of 1983. During the intervening years one of the main aims of the series has remained unchanged, i.e. to publish in each volume a number of chapters—considerably more extensive than papers in technical journals, but shorter than monographs—each giving an authoritative and comprehensive up-to-date review of the state-of-the-art of a subject area within hydraulic engineering.

The first two volumes were oriented each broadly towards a main theme without being too narrow. Volume 1 dealt with the role and some applications of computational hydraulics in hydraulic structures design and with irrigation structures and related sediment problems; the hydraulic structures orientation continued in the second volume concentrating on vibration and operation of structures and on the design of spillways and energy dissipation. The third volume saw a change of emphasis, with seven chapters ranging over a broader field of river, estuarine and coastal hydraulics including dispersion, flood routing, ice engineering, inland waterways and port design. Although the fourth volume dealt predominantly in four chapters with groundwater hydraulics and development, two further contributions on lake hydraulics and tidal power—both of substantial topicality—were included.

Although, in the long term, it is likely that each future volume may deal with a more clearly defined single main theme within hydraulic engineering, for the time being the editor is glad to have secured—for this and the next volume—the cooperation of a number of authors well known in their subject area, who have written original contributions, each with a comprehensive list of references.

There is one further innovation in this fifth volume. Previously the number of chapters varied between five and seven; in this volume, due to the need for a more thorough treatment of the subject matter, particularly in two of the contributions, only four chapters are included, as otherwise Volume 5 would greatly exceed its planned length.

The first chapter, on water power development, deals in its first three sections with some general matters of hydropower utilisation, but the bulk of the text is oriented towards all aspects of low-head river developments, including their layout, planning criteria, magnitude and evaluation of producible power and energy, generating sets and their operation, electrical equipment, environmental aspects and economic appraisal. A special section on small size hydropower plants is also included. The text is well documented by figures and contains both the theoretical background as well as practical advice drawing on the author's lifelong involvement in hydropower development. Although some of the material has been published by the author before, it is dispersed in many publications; this text has been thoroughly revised and updated and contains a great deal of previously unpublished work. The first part of the text forms also an introduction to a future text by Professor Mosonyi on high-head power development, which is likely to be included in the next volume.

Developing further the previous treatment of ice engineering, the next chapter, on intake design for ice conditions, discusses one topic in ice engineering which is of paramount importance in power development and water supply in general. It deals with the characteristics of ice problems that must be taken into account in the design and operation of intakes, the ice growth and production rates and its behaviour at intakes in various conditions and with the counter measures to be taken. The chapter concludes with some typical case studies illustrating the magnitude and diversity of the ice conditions at intakes.

Although, in Volume 3 of this series, some aspects of wave climate were briefly mentioned in the chapter on ports, and the dispersion in coastal waters was extensively covered, many aspects of the sea face of estuaries, particularly those referring to tidal currents and waves and their effect on sediment transport, had not been previously discussed. This is an area with many new, exciting, recent developments in analysis and computation; starting with a description of three typical estuaries with widely differing characteristics (Santos in Brazil, Rio Pungue in Mozambique and the Mersey in the UK), the author discusses the methods of study of the hydraulic behaviour of estuaries, concentrating on the numerical analysis of interactions between estuaries and seas. The detailed treatment of regional modelling of tidal flows and of tidal currents at estuary mouths, as well as of regional modelling of waves and of nearshore waves and currents, is followed by the discussion of sea bed stresses and induced sediment transport and a summary of the properties of various wave models. The chapter concludes with a section speculating how the previously mentioned three estuaries could be modelled using recently developed techniques.

The last chapter of this volume looks to one specific development in water resources engineering of lowlands—the polders. The Dutch engineers have, of course, long been the pioneers in this area, and three Dutch authors, after a brief description of polders, deal in some detail with the soil types encountered in polders and particularly with the complex water management of polders, including design of irrigation and drainage and flood control systems. The text concludes with a look to the polders of the future.

In presenting this fifth volume of the series, the Editor trusts that its individual texts as well as the volume as a whole will be received as a further contribution towards progress both in the specialised fields presented here and in hydraulic engineering in general.

P. Novak

CONTENTS

LIST OF CONTRIBUTORS

G. D. ASHTON
   US Army Cold Regions Research and Engineering Laboratory,
   Hanover, New Hampshire 03755, USA

J. LUIJENDIJK
   International Institute for Hydraulic and Environmental Engineering
   (IHE), Delft, The Netherlands

D. M. MCDOWELL
   Emeritus Professor of Civil Engineering, University of Manchester,
   UK. Present address: *13 Powis Villas, Brighton BN1 3HD, UK*

E. MOSONYI
   Institut für Wasserbau und Kulturtechnik, Universität Karlsruhe,
   Kaiserstrasse 12, D-7500 Karlsruhe 1, Federal Republic of Germany

E. SCHULTZ
   IJsselmeerpolders Development Authority, Postbus 600, 8200 AP
   Lelystad, The Netherlands

W. A. SEGEREN
   International Institute for Hydraulic and Environmental Engineering
   (IHE), Delft, The Netherlands

Chapter 1

WATER POWER DEVELOPMENT: LOW-HEAD HYDROPOWER UTILIZATION

E. MOSONYI

*Institut für Wasserbau und Kulturtechnik,
University of Karlsruhe, Federal Republic of Germany*

1 INTRODUCTION

The harnessing of water power resources had been abandoned or at least decelerated for a few decades throughout the world after the prophecies of the 1950s: (a) that oil would be available in sufficient quantities to cover all requirements and extremely cheap for a long time, and (b) that nuclear power plants would very soon cover all the world's further energy demands without risk. It turned out, however, that both of these assumptions were no more than naive expectations propagated unfortunately even by scientists and professional engineers. Consequently, in some countries, the planning and implementation of several promising hydroelectric projects has been delayed or completely stopped for about two decades.

The author deems it unnecessary to discuss the above mentioned Utopian dreams; facts have in the meantime justified the opinion of numerous hydraulic engineers concerning the real value of water power.

Nevertheless, new obstacles to progress in hydropower utilization have recently emerged because of its environmental impact. It cannot be denied that several defects, even irreparable failures, occurred owing to ignorance or negligence of environmental requirements. Overhasty planning and

1

construction carried out under political pressure has also been the reason for failure. On the other hand, unjustifiable attacks, even on schemes planned with the utmost care for environmental protection, are becoming more frequent in some countries. It is, however, the author's firm conviction that recent progress in research and up-to-date technology permits, in most cases, the selection of a proper design and of measures compensating for or diminishing environmental impacts to acceptable limits. Hence it can be presumed that the construction and operation of hydropower projects entails, in general, disadvantageous environmental effects to a lesser degree than that of any other type of power generating plants.

The competitiveness of hydroelectric plants has been significantly enhanced during the last decades by progress in various fields of technology both in civil and mechanical engineering, e.g. in new foundation and grouting methods, improved tunnelling techniques, lightweight and thus less expensive steel structures, improved and new designs of gates, turbines and generators, simplified superstructures, etc. A more efficient technology permits a considerably shortened construction time in comparison with earlier schedules, with substantial economic consequences.

It has to be borne in mind that preliminary studies, planning, design, construction and installation, operation and maintenance of hydroelectric schemes are all of multi-disciplinary character. Therefore stability, safety, operational reliability and economic effectiveness of hydropower plants strongly depend on achievements in research in several fields such as hydrology and meteorology, geology and soil mechanics, hydromechanics of civil structures and machines, structural analysis and electrotechnics. Especially research on vibration, cavitation, perfected instrumentation and measurement methods in scale-model tests and new procedures of regional hydrological analyses have contributed markedly to the improvement of plant designs. In addition, it must be mentioned that a sound project evaluation may require, according to the specific features of the case, a high competence in various other domains (biology, ecology, sociology, economics, etc.)

Some remarks concerning the content of this chapter have to be made although typical general layouts are displayed. Structures which may be a part of any other scheme except hydropower plants are not further discussed in detail (e.g. dams, weirs, ship locks, fish passes, intakes). A concise discussion of the turbines and electrical equipment is essential, since the power plant planner is not able to elaborate even a preliminary sketch of the plant without making an estimate of the number and type of generating units and of approximate operational characteristics (specific speed, rated speed) and, consequently, of the elevation and turbine runner diameter. It cannot be sufficiently emphasized that the general arrangement of the entire powerhouse completely depends on the latter two parameters.

Considering the fact that only about 10–20% of the economically feasible water power resources of our world have been utilized up till now, it can undoubtedly be stated that hydropower engineers may have still much confidence in their future. The reason for the seeming uncertainty about the degree of exploitation indicated above can be explained by the fact that the assessment of the quantity to which the harnessed power is related, i.e. the economically feasible potential, continuously changes according to technological progress and policy in economic evaluation.

The terms project, scheme and development are used as synonyms for plants, comprising all structures erected and measures applied for utilizing a selected water power resource. Water power, hydropower and hydroelectric plants are used as synonyms too. The word station refers to the coherent complex of powerhouse and weir (spillway), while a river barrage includes also ship lock(s). The cascade or chain of river barrages rendering possible or improving navigation results in a canalization of the river. The impounded headwater stretches are alternatively called reaches or ponds. Turbine and wheel are applied synonymously. The terms unit, aggregate and set are synonymously collective designations for the turbine and its generator, including, if any, the gear drive too.

Data in captions of figures showing projects and powerhouses respectively are given in the following way: name, river, country, year of commissioning, plant discharge capacity (plant design flow), rated net head (or in some instances the available head range), rated (installed) capacity. This set of data, however, is not complete in every case. A certain discrepancy in indication of the head could not be avoided because in some publications only the gross head has been referred to.

2 SOURCES OF MECHANICAL ENERGY IN WATER

2.1 Physical Principles of Water Power Utilization

In order to obtain a clear picture of the practical possibilities for extracting power from the natural potential of the watercourses, we must know what happens to the energy inherent in natural rivers. Particles of flowing water possess potential energy depending on the altitude and kinetic energy depending on the changes in the flow velocity. Although kinetic energy of rivers arises at the expense of potential energy, it is insignificant compared with the dissipated potential energy which is used to overcome the friction of the turbulent flow, to supply energy to whirls, to scour the river bed and to transport sediment. This mechanical work is finally completely converted into heat and is lost once and for all.[1]

Thus the fundamental principle of hydropower utilization is to reduce the dissipation of energy. Consequently, a certain part of the potential power appearing in utilizable head can be regained from nature. Friction losses can be diminished basically by reducing the velocities by constructing a dam or by reducing the head required for the conveyance of water by diverting the whole or a part of the flow into a by-pass canal, or finally by a combination of the two methods.

Thus it depends mainly on topographical and fluvial-morphological conditions to what degree the natural potential expressed in head can be harnessed from a technological point of view. One has to bear in mind of course the various restricting conditions of the region (anthropogenic development) and economic feasibility.

2.2 Units

In the field of hydroelectric development, power and capacity are usually given in kilowatts (kW) or megawatts (MW), and the generated and producible energy in kilowatt-hours (kWh) or megawatt-hours (MWh) or, in cases of high productivity, gigawatt-hours (GWh). Formerly it was also usual to state turbine capacities or even plant capacities in horsepower units. It has to be noted that there is a slight difference between the metric horsepower (HP) and the British horsepower (hp):

$$1\,\text{HP} = 0.9862\,\text{hp} \quad \text{or} \quad 1\,\text{hp} = 1.014\,\text{HP} \tag{1}$$

Since mechanical energy (work) is usually expressed in newton-metres (Nm) or kilonewton-metres (kNm) and the power (capacity) in newton-metres per second (Nm/s) or kilonewton-metres per second (kNm/s) it is necessary to introduce the relevant conversion factors.[2,3]

As (by SI definition)

$$1\,\text{joule (J)} = 1\,\text{Nm} \tag{2}$$

and

$$1\,\text{watt (W)} = 1\,\text{Nm/s} \tag{3}$$

it follows that

$$1\,kW = 1000\,Nm/s = 1\,kNm/s \text{ (power)} \tag{4}$$

and

$$1\,kWh = 1000\,Nm/s \times 3600\,s = 3\cdot6 \times 10^6\,Nm$$
$$= 3600\,kNm \text{ (energy or work)} \tag{5}$$

or, when comparing eqn (2) with eqn (5),

$$1\,kWh = 3\cdot6 \times 10^6\,joules = 3\cdot6 \text{ megajoules (MJ)} \tag{6}$$

Since we are still in a transition period towards the universal usage of accepted technological measures (kW and kWh) and SI units, and have to evaluate data from old plants, it is desirable to recapitulate also older units for energy (work) (kg-force m or kpm) and power (horsepower):

$$1\,HP = 736\,W = 0\cdot736\,kW \tag{7a}$$

$$1\,hp = 746\,W = 0\cdot746\,kW \tag{7b}$$

$$1\,kWh = 367\,000 \text{ kg-force} \times m = 367\,000\,kp \times m \tag{8}$$

In connection with heat production and heat losses, the calorie (cal) and kilocalorie still occur in publications:

$$1 \text{ calorie (cal)} = 4\cdot187 \text{ joules (J)} \tag{9}$$

$$1 \text{ kilocalorie (kcal)} = 4\cdot187 \text{ kilojoules (kJ)}$$
$$= 1\cdot163 \text{ watt-hours (Wh)} \tag{10}$$

and consequently

$$1\,kWh = 860\,kcal \tag{11}$$

To indicate the revolving speed of machines the 'revolution per minute' (rpm) is commonly used, its dimension being $1/min$. The frequency of vibration, cycles, waves or impulses is mostly related to the second; thus, expressed in hertz units (SI equivalents):

$$1 \text{ rev. or cycle per second (cps)} = 1 \text{ hertz (Hz)}$$
$$= 1\,s^{-1} \tag{12}$$

and accordingly

$$1 \text{ rpm } (min^{-1}) = 1/60\,Hz = 0\cdot01667\,cps\,(s^{-1}) \tag{13}$$

2.3 Power in Flowing Water

When assuming uniform steady flow between two cross-sections of a watercourse, the power represented by H metres difference in water surface elevation between two sections and which dissipates along the stretch L can for a flow of $Q\,\text{m}^3/\text{s}$ be expressed as

$$P = \gamma Q\left(H + \frac{v_1^2 - v_2^2}{2g}\right) [\text{Nm/s}] \qquad (14)$$

where v_1 and v_2 are the mean velocities in the two sections. Neglecting the usually slight difference in the kinetic energy and substituting for $\gamma = \rho_g = 9810\,\text{N/m}^3$, we obtain

$$P = 9810QH\,[\text{Nm/s}] \qquad (15)$$

or from eqn (4) (for Q in m^3/s and H in m)

$$P = 9\cdot81QH\,[\text{kW}] \qquad (16)$$

Equations (14)–(16) give the theoretical (physical or potential) power of the selected river stretch at a specified discharge.

A watercourse consists of stretches between major tributaries and/or outflows, which can be characterized, during a specified state of flow, by constant or averaged discharges. Accordingly the theoretical power resource of any river or a river system, at the selected state of flow is

$$P = 9\cdot81\,\Sigma Q_i H_i\,[\text{kW}] \qquad (17)$$

where Q_i and H_i are the values pertaining to stretch i.

When evaluating integrated power resources for an entire watercourse or drainage basin (river system) there arises the question of what values of H and Q should be used in the calculation. If we disregard very short river stretches, the change of H with stage is negligible; on the other hand, a thorough examination of the discharge variation is of great importance.

Potential power is evaluated to obtain an inventory of possibilities and for ranking the feasible power sites. Therefore diverse magnitude of the potential hydropower resources are usually determined on the basis of geodesical and hydrological measurements:

1. Minimum potential power is based on the smallest runoff, having a duration of 100% ($P_{p100\%}$). This value is usually of small interest.
2. Small potential power is calculated from the 95% duration discharge ($P_{p95\%}$).
3. Median or average potential power is gained from the 50% duration

discharge ($P_{p50\%}$).
4. Mean potential power results by evaluating the annual mean runoff (P_p).

The sum of these values for an entire stream is termed gross river power potential (Economic Commission for Europe (ECE), Committee on Electric Power).[4]

Since it is not economically feasible to harness floods, and therefore the utilization for every site is usually limited to an individually selected plant discharge capacity, there is no reason for including the entire magnitude of peak flows in the inventory either of the potential power or of the potential annual energy. Therefore, the annual discharge duration curve will be truncated at a certain t days duration of discharge, which can be simply the median ($\sim$ 182 days or 50% duration, denoted by Q_{182} or $Q_{50\%}$), or a higher Q_t ($t < 182$ days) can be selected by specialists who are familiar with the local conditions and future plans for power supply. Accordingly the annual magnitude of potential (theoretical) energy can be computed in kWh (Fig. 1):

$$E_p = 24 \times 9 \cdot 81 H \left(Q_t + \sum_{t}^{365} Q_i\right) \simeq 235 HA \,[\text{kWh}] \qquad (18)$$

where Q_i denotes the daily mean flow during the period $365 - t$ and A the hatched area cut by Q_t. (Dimensions kNh/(m$^3 \times$ day/s) pertain to the factor 235, while A is read in m$^3 \times$ day/s.)

Neither all potential power nor potential energy is available for direct

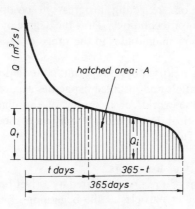

FIG. 1. Annual discharge curve.

utilization. Only a part of the geodetically evaluated difference can be regained by providing a concentrated drop in the water surface, the so-called head available for the turbines (see Section 2.1), and for many physical reasons and because of anthropogenic influences not every river stretch can be earmarked for full or even partial utilization. Besides, there are instances where, because of demands for water with higher priority than power generation, the natural discharges are not fully available for the power station. Accordingly, the duration curve has to be modified to obtain the reduced discharges utilizable for the turbines. With diversion type projects, generally also a predetermined permanent duty flow has to be provided for in the river stretch by-passed by the power canal. Thus, after having selected a power site, the producible power, at any available gross head and discharge, can be established as (see eqn (16))

$$P = c_0 QH [\text{kW}] \qquad (19)$$

where all head losses and the constant 9·81 are integrated in the overall power coefficient of the plant c_0 which, according to the individual technical solution, shows a variation roughly between 7 and 8·5.

Thus the technically available resources and the practically feasible power estimates can be established. A further question is which of the envisaged schemes can also be accepted as economically promising ones and, if so, at what discharge capacities. This evaluation, however, changes in the course of time, since it greatly depends on regional or even international technological power development and on the power policy.

The plant discharge capacity, also termed plant design flow, has undergone a remarkable change during the last 100 years.[5] Figure 2 demonstrates this evolution, reflecting progress in co-operation of power suppliers and in power technology. Thus the degree of utilization at some modern plants attains magnitudes in the range of $Q_{25\%}$ to $Q_{15\%}$ (even up to ∼ 12%).

The scope of this treatise does not permit either a detailed presentation of theoretical potentials and installed capacities according to river basins and countries in the world, or a survey of the changes of economic estimates and historical development;[6,7] only an earlier summary by continents is given in Table 1.

2.4 Energy Capacity of Reservoirs

If a stream flows out of a reservoir created by a river barrage or high dam, the energy of all water particles at the beginning of emptying is H_0, i.e. height of water level in the reservoir in relation to a selected datum level.

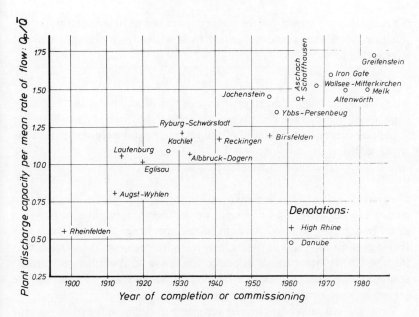

FIG. 2. Examples revealing the progress in degree of utilization. (Compilation by the author.)

TABLE 1

ESTIMATE OF HYDROPOWER EXPLOITATION IN THE WORLD[a]

Continent	Capacity in GW (1000 MW)		Rate of exploitation (%)
	Available	Exploited (installed)	
Europe	290	190	65
North and Central America	375	105	28
South America	690	15	2
Asia	1350	45	3
Africa	1100	13	1
Australia and Oceania	150	10	7
World	3955	378	10

[a] Based on data compiled in the 1970s; the installed capacities may be, at present, about 5–20% higher.

Thus the inherent energy of the water masses stored in a reservoir, i.e. the potential work that could be gained by completely emptying it without losses, can be expressed as

$$E = \gamma V H_c \, [\text{Nm}] \tag{20}$$

where H_c is the height pertaining to the centre of gravity of the V storage capacity of the reservoir. From eqn (5) and by substituting $9810 \, \text{N/m}^3$ for γ, the physical energy potential of the reservoir can be converted into the usual technical unit:

$$E = V H_c / 367 \, [\text{kWh}] \tag{21}$$

Obviously, owing to the various kinds of losses, only part of this physical energy can be converted into useful work. The varying physical power is given by eqn (16), where Q depends on the opening of the outlet gates and turbines and H pertains to the water level in the reservoir at any particular instant. On the other hand, in periods of inflow to the full reservoir and assuming that the outflow can be kept in balance with the inflow, the power is permanently related to the head H_0.

Reservoirs can be characterized according to height of the dam,[8] storage capacity,[8] created lake surface or inherent physical energy, according to eqn (21).

According to a most up-to-date publication[8] the world's highest dams, of 200 m or more, and the world's largest reservoirs, equalling or surmounting $50 \times 10^9 \, \text{m}^3$ (50 milliard m^3; according to English usage 50 billion m^3; according to SI definition 50 giga-cm^3), are listed in Tables 2 and 3.

3 CLASSIFICATION OF HYDROPOWER UTILIZATION

3.1 Harnessing of Watercourses (Traditional Water Power Development)
The theoretical (physical) potential of all watercourses of the world evaluated according to eqns (17) and (18) has been estimated by various authors and institutions during the last decades.[6,9–11] The assessment of the world's theoretical hydropower capacity (related to the annual mean discharges) is around 5·6 million MW, while estimation of the so-called installable (available) capacity varies between 2·3 million and 3·8 million MW (3·8 TW).

The theoretical annual energy content of the world's rivers has earlier been estimated at 36 000 TWh (1 terawatt-hour = 1×10^9 kWh); never-

TABLE 2
WORLD'S HIGHEST DAMS
(after T. W. Mermel)

Name	Country	River	Dam type[b]	Height (m)	Year of commissioning
Rogun	USSR	Vakhsh	E/R	335	1985
Nurek	USSR	Vakhsh	E	300	1980
Grand-Dixence	Switzerland	Dixence	G	285	1962
Inguri	USSR	Inguri	A	272	1980
Boruca[a]	Costa Rica	Terraba	R	267	1990
Vaiont	Italy	Vaiont	A	262	1961
Chicoasén	Mexico	Grijalva	E/R	261	1980
Tehri[a]	India	Bhagirathi	E/R	261	1990
Alvaro Obregon	Mexico	Tenasco	G	260	1946
Kishau	India	Tons	E/R	253	1985
Sayano-Shushensk	USSR	Yenisei	G/A	245	1980
Guavio[a]	Colombia	Guavio	E/R	243	1987
Mica	Canada	Columbia	E/R	242	1973
Mavoisin	Switzerland	Drange de Bagnes	A	237	1957
Chivor	Colombia	Bata	R	237	1975
Chirkey	USSR	Sulak	A	233	1978
Oroville	USA	Feather	E	230	1968
Bhakra	India	Sutlej	G	226	1963
El Cajón	Honduras	Humuja	A	226	1985
Hoover	USA	Colorado	A/G	221	1936
Contra	Switzerland	Verzasca	A	220	1965
Mratinje	Yugoslavia	Piva	A	220	1976
Dworshak	USA	Clearwater N.Fork	G	219	1973
Glen Canyon	USA	Colorado	A	216	1966
Taktogul	USSR	Naryn	G	215	1978
Klamm[a]	Austria	Dorferbach-Matrei	A	220	1989
Daniel Johnson	Canada	Manicouagan	M	214	1968
Dez	Iran	Dez	A	213	1962
Keba	Turkey	Euphrates	E/G/R	207	1974
Khudoni[a]	USSR	Inguri	A	201	1990
Kölnbrein	Austria	Malta	A	200	1977
Lower Tunguska[a]	USSR	L. Tunguska	E/G	200	1994
Luzzone	Switzerland	Brennodi L.	A	208	1963
San Roque[a]	Philippines	Agno	E/R	210	1992
Almendra	Spain	Tormes-Duero	A	202	1970
Karun	Iran	Karun	A	200	1976

[a] Planned or under construction. [b] A = arch, E = earth, G = gravity, R = rockfill.

theless, later assessments reveal values from 44 000 up to almost 50 000 TWh/a. The available energy appears in recent inventories on hydropower resources at about 10 000 TWh/a.

The estimate of the installable power and, consequently, of producible energy varies greatly with time, showing an unmistakable increase over the

TABLE 3
WORLD'S LARGEST-CAPACITY RESERVOIRS
(after T. W. Mermel)

Name	Country	Capacity $(10^9 \, m^3)$
Owen Falls	Uganda	2 700
Kahkovskaya	USSR	182
Bratsk	USSR	169
Aswan (High)	Egypt	169
Kariba	Zimbabwe	160
Akosombo	Ghana	148
Daniel Johnson	Canada	142
Guri	Venezuela	138
Kama	USSR	122
Bennett W.A.C.	Canada	74
Krasnoyarsk	USSR	73
Zeya	USSR	68
Cabora Bassa	Mozambique	63
La Grande 2	Canada	62
La Grande 3	Canada	60
Ust Ilim	USSR	59
Volga-V.I.Lenin	USSR	58
São Felix	Brazil	55
Caniapiscau	Canada	54
Cerros Colorados	Argentina	54
Chapeton[a]	Argentina	54
Shintoyone	Japan	54
Upper Wainganga[a]	India	51
Bukhtarma	USSR	50

[a] Planned or under construction.

last decades. The tendency of an almost constant growth of available hydropower resources is partly the outcome of technological progress, but even more so a consequence of the disillusionment about 'extremely cheap oil' and of disappointment about the too optimisitic assumptions on the energy production costs of other types of power plants.

In the early 1970s about 10% of the available hydropower resources had been harnessed. For the end of 1985, a quantity of 2200 TWh/a producible hydroenergy was anticipated which corresponds to about a 20% rate of exploitation. Summary data on recent developments in hydropower[7] and on the largest-capacity projects of the world are published. Table 4 lists the plants achieving or having a planned capacity of 2500 MW or more.[8] The participation of hydroelectricity in world production of electric energy

TABLE 4
LARGEST-CAPACITY HYDROPOWER PLANTS OF THE WORLD (1985)
(after T. W. Mermel)

Name	Country	Rated capacity (MW)	
		1985	Planned
Itaipu	Brazil/Paraguay	2 800	12 600
Guri	Venezuela	2 800	10 000
Tucurui	Brazil	3 960	8 000
Grand Coulee	USA	6 494	6 494
Sayano-Shushensk	USSR	6 400	6 400
Corpus[a]	Argentina/Paraguay	–	6 000
Krasnoyarsk	USSR	6 000	6 000
La Grande	Canada	2 000	5 328
Churchill Falls	Canada	5 225	5 225
Bratsk	USSR	4 500	4 500
Ust-Ilim	USSR	3 675	4 500
Yacyretá-Apipe[a]	Argentina/Paraguay	–	4 050
Cabora Bassa	Mozambique	2 000	4 000
Rogun[a]	USSR	–	3 600
Paulo Afonso I	Brazil	1 524	3 409
Ilha Solteira	Brazil	3 200	3 200
Gezhouba	P.R. of China	2 715	2 715
John Day	USA	2 160	2 700
Nurek	USSR	900	2 700
Revelstoke	Canada	900	2 700
São Simão	Brazil	2 680	2 680
Mica	Canada	1 730	2 610
Volgograd	USSR	2 563	2 563
Itaparica[a]	Brazil	–	2 500

[a] Planned or under construction.

shows a diminishing trend: in 1950 it was 35·8%, in 1974 it decreased to 23·0%, in 1985 to 18·4%, and 14% is anticipated for 2000.

3.2 Pumped Storage

Pumped storage plants do not produce additional energy for the network; their operation is a double-phase power conversion accomplished by recycling of water masses between two reservoirs. The originally unique objective was to store the surplus energy producible by any type of power plant in the off-peak periods of the power system, in the form of hydraulic potential energy, in order to regain it during periods when the peak demand on the system exceeds the total capacity of the co-operating stations. Thus a pumped storage project consists of the following main structures: (a) lower basin; (b) power plant which operates as a pumping station during the off-peak periods and generates power in the peak-load times; (c) penstock(s) or pressure shaft(s); (d) upper basin with intake headwork (Fig. 3).

Coverage of daily, weekly and seasonal (annual) peak demands can be differentiated according to the selected recycling period. In the earliest stages of development the pumped storage plants were almost exclusively designed to carry the daily peak loads, and this is still their most common mode of operation. Up to the 1960s, with few exceptions, the powerhouses were equipped with three-machine aggregates (pump, motor/generator, turbine), while in the last decade the achievements in constructing high-efficiency pump-turbines (reversible hydraulic machines) resulted in the increasing application of two-machine sets (motor/generator and pump/turbine).

Experience with pumped storage has shown that, apart from the coverage of peak power demands, it can considerably contribute to the efficiency and safety of the power system by some special modes of

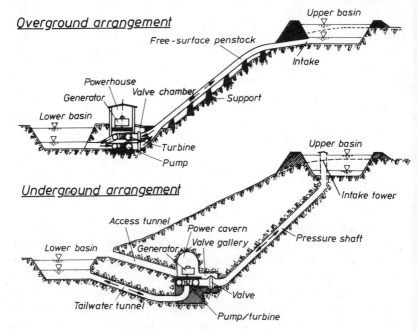

FIG. 3. Conceptual sketch of pumped storage plant.

operation, i.e. by participation in or guidance of the network frequency control, provision for regular and emergency power reserve and improvement of the network's power factor (by synchronous condenser or induction motor operation).

Pumped storage can also be extended to multi-purpose schemes combining the above described power management with irrigation, potable and industrial water supply, low-water enhancement for environmental quality control, flood alleviation and recreation and landscape protection, etc.

Extensive literature is available on pumped storage and its environmental effects. The proceedings of the international conference on this topic at the University of Wisconsin, Milwaukee, September 1971, apart from presenting the geotechnical, technological, economical and environmental aspects of planning, design, construction and operation of both single-purpose and multi-purpose pumped storage schemes, contain an exhaustive bibliography.[12] Further references to pumped storage will be given in a future chapter dealing with high-head developments.

3.3 Tidal Power Generation

In the present stage of technological progress, the utilization of tides for power generation appears to be most promising (as manifested by profound multi-disciplinary studies).[13] The tidal amplitude attains considerable magnitudes in certain oceanic regions, e.g. at an Atlantic coastal stretch of Canada 13·5 m (15 m), in the Bristol Channel 10 m (14 m), on the French Atlantic coast 8 m (13·5 m), where the first values denote the mean annual ranges and the figures in brackets refer to the maximum, i.e. equinoctial amplitudes. In the Pacific region high tides are also recorded, e.g. along the coasts of China and the USSR, where the utilization of mean annual amplitudes of 6–9 m have been considered. Recently Wilson and Balls[14] presented in these volumes a comprehensive discussion of the present state of the art in tidal power, and it is worthwhile recording here only that the first high-capacity tidal project in the world, the Rance plant on the French Atlantic coast equipped with twenty-four 10 MW capacity bulb tubular generating sets, was commissioned in 1966.[15,16] Although originally economists and engineers doubted the economic feasibility of this tidal plant, it has been accepted as successful since its energy production cost makes it competitive with other power resources in the French power system. This success and further technological developments[14,17] have stimulated the construction and planning of tidal plants in Canada,[18] the USSR, China and the UK.[14]

3.4 Wave Energy Conversion

The average power transmitted by a sinusoidal wave on a length of 1 m is

$$P_0 = \frac{1}{1000} \frac{\gamma g}{32\pi} H^2 T \left\{ 1 + \frac{2kd}{\sin [h(2kd)]} \right\} \tan [h(kd)] \, [\text{kW/m}] \quad (22)$$

where γ is the specific weight of the water $(9810 \, \text{N/m}^3)$, $g = 9.81 \, \text{m/s}^2$, H = wave height in m, T = wave period in s, d = water depth in m, k is the wave number expressed as

$$k = \frac{2\pi}{\text{wavelength in m}}$$

This theoretical equation related to a single wave has to be corrected to obtain values corresponding to a real wave spectrum.

By substituting the constants into eqn (22) and applying some approximations, the physical (potential) specific wave power may be expressed as

$$P_0 = 0.995 \, H^2 T \simeq H^2 T \, [\text{kW/m}] \quad (23)$$

It is assumed that about 50% of potential wave energy can be captured and this convertible power can be further utilized with an efficiency of around 50% (generation and transmission). Grove-Palmers estimated the electric power producible along a 1000 km long coastline of the UK as high as 12 000 MW.

The wave energy converter generally comprises the following main parts: (1) reference frame (body); (2) mechanical power conversion device; (3) electrical equipment; (4) power transmission; (5) structural linkage between converter and sea bed.

Several types of converter systems have been studied and proposed for industrial use. Some are floating devices, while others are rigidly fixed to the sea bed. Since the technology of wave power utilization is still *in statu nascendi*, and there is still no general agreement as to which type from the innumerable proposals will provide the most economical solution, only a brief summary of some of the devices is given here. Further details are available in the literature.[19–21]

The main categories, named according to the principle of their functioning, are as follows:

(a) The terminator is a floating device positioned perpendicular to the direction of the predominating wave flux. The converter incorporates flexible air bags which, pendulated by waves, induce two-direction air currents in a self-rectifying air turbine, the latter coupled to a generator (Salter's Duck, Sea Clam).

(b) The attenuator penetrates into the wave, its axis being parallel to the direction of wave propagation (Cockerell Raft, Lancaster Flexible Bag, etc.). Thus the power of the wave is absorbed progressively by the device's elements as the wave crest moves along its length.

(c) The Russel Rectifier consists of a series of chambers, each housing a turbine.

(d) NEL's Oscillating Water Column is rigidly fixed to the sea bed. Either a membrane-closed or an open oscillating water column responds to the oncoming wave and induces oscillations of air masses which drive the turbines. Figure 4 elucidates the operation of the rectifying values resulting in a unidirectional airflux through the turbine.

(e) The Buoy Power Converter is, in contrast to all the above listed 'linear' power converters, a 'point absorber'. The spherical buoy with an opening at its lower end is submitted to a heaving motion by the waves generating an air current (Norwegian solution), or accelerated water masses (Swedish variant) driving the turbine.

The recently published results of the UK research programme do not at present indicate promising economical results.[22]

3.5 Depression (Solar) Schemes

This possibility of hydropower generation is based on the high potential evaporation in hot-climate regions with sea water being conveyed through

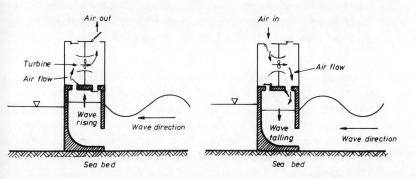

FIG. 4. The NEL breakwater wave energy converter. This drawing elucidates the rectifying concept applied similarly in various other types of converters. (After G.W. Moody and G. Elliot.)[19]

a canal or tunnel into a basin below sea level from where it evaporates. The level to which the depression fills will be determined by the equilibrium between inflow and evaporation. The relation between the rate of flow Q (m³/s) and evaporating surface A (m²) and the annual evaporation h (m/a) can be obtained as[23]

$$Q = \frac{Ah}{31 \cdot 5 \times 10^6} \; [\text{m}^3/\text{s}] \tag{24}$$

The available head for utilization after reductions due to the conveyance losses pertaining to any discharge is given by the difference between sea and steady-state lake level. Since a higher discharge requires a greater evaporation surface (and vice versa), it follows from the above that the utilizable discharge is in inverse relation to the attainable head. The powerhouse suitably located at the bank of the (depressed) lake houses the generating aggregates.

Up to the present, due to the high investment costs, no scheme of this type has been implemented although several deep natural basins situated in the vicinity of sea coasts were already identified some decades ago. Two solar projects have been intensively studied and elaborated as construction-ripe plans:

— The Qattara depression plant in Egypt, to be created by a 75 km long connection (deep-cut canal and/or tunnel) between the Mediterranean and the depression to be filled to provide an evaporation surface of 20 000 km²; designed for 660 m³/s discharge and 40 m head.[24]

— The Dead Sea scheme in Israel with an inlet from the Mediterranean through an 80 km long connection, providing a head of 390 m; the planned plant capacity lies between 100 and 300 MW.[25]

3.6 Classification of Traditional Hydropower Projects According to Objectives and Main Physical and Technological Parameters

(1) The turbines can (a) *directly drive pumps* or other types of machines (in factories, mines, etc.), or (b) be *coupled to generators* thus producing electric power. The former solution belongs more to the past; its application at present is very exceptional. (Therefore this treatise deals with the hydroelectric mode of utilization only.)

(2) *Single-purpose* and *multi-purpose schemes*. The possible secondary objects are: flood control, irrigation, navigation, industrial and/or municipal water supply, low-water enhancement, groundwater recharge,

drainage, provision for recreation and sporting facilities, improvement of environmental qualities. In several instances priority is given to one or more of the above purposes and power generation may be a secondary aim only.

(3) According to the general layout of the entire project the following main distinctions are made: the power station proper is located (a) *in the river*, (b) in or at one end of a *diversion* (canal, penstock, tunnel).

(4) According to storage the following differentiation is usual. If the dam or river barrage is very low and the topographical and/or operational constraints allow no or only an insignificant change in the headwater level the natural discharge Q is utilized by the turbines up to the plant design flow (plant discharge capacity, plant rated discharge) Q_p, while during periods when $Q > Q_p$ the excess flow $Q - Q_p$ has to be released through the weir (over the spillway). This type of project is termed *run-of-river plant*. When the available storage capacity and the permissible change in headwater elevation render possible a small-scale discharge control so that the plant can contribute to the daily peak power production, the term *pondage* is used. Reservoirs with relatively (to the integrated natural flow) large capacities can provide (a) for generation of weekly, seasonal or annual peak energy, or (b) for partial or complete equalization of long period inflows. Thus weekly, seasonal, yearly and multi-annual *storages* can be distinguished.

(5) The attributes *open-air* and *underground* solution refer to the location of the powerhouse (machine hall) of the project.

(6) Plants are sometimes characterized according to their installed power capacities: *high capacity, medium capacity, low capacity*, etc. Such specification, of course, has not proved realistic, for this type of valuation strongly depends on the existing hydropower site potentials in the various river basins and countries. During recent years extraordinary world-wide interest has been focused on the so-called '*small hydro*'. This type of plant (also named small-scale or mini plant) cannot be defined unequivocally, i.e. with a fixed power limit, either. However, a separate analysis of the very-low-power plants is justified because they require specific treatment in design and implementation. According to diverse suggestions the capacity of typical small plants could be limited by 1 to 5 or even 15 MW. (It is not the capacity but the specific features, if any, of the small plant that matter.)

(7) According to the order of magnitude of the utilizable net head for the turbines it is customary to differentiate between *low-head and high-head plants*. (Previously some authors suggested a three-stage grouping by

discerning medium-head developments too.) This presentation is based on the separation of low-head and high-head projects, in spite of the fact that it is difficult to define an unequivocal limit with a fixed figure concerning the head. Therefore, it seems to be more appropriate to describe the main characteristics of the two types.

Low-head plants are equipped with axial turbines or, at the present rarely, with high-speed Francis turbines, thus utilizing heads generally up to around 30 m, less frequently up to around 50 m. With low-head stations the intake, the machine hall and the conveyance system in between are usually integrated in a single structure—the powerhouse.

High-head power stations usually house Francis turbines or Pelton wheels and, owing to the higher heads, the intake, the conveyance (penstock or shaft) and the powerhouse are usually separate structures.

3.7 Classification of Plants According to their Role in the Power Supply System

The operation mode of a hydropower plant depends partly on the available input, i.e. the natural flow conditions and storage possibilities, and partly on the output demands, i.e. the consumers' needs.

(a) *Isolated* plants, located mostly in remote regions, are not connected to a regional or national power network; they provide electric energy for a lonely consumer or for a tiny consumer group. A *co-operating* plant supplies its power into a regional, national or multi-national grid and, consequently, its operation mode, beside the physical potential input, is greatly influenced by both the consumers' demands and the power available from all other plants connected to the network (i.e. other hydropower projects, thermal and nuclear stations). 'Available' refers here not only to the capacities but includes also economical operation.

(b) According to the kind of energy fed into the network the plants can be sorted roughly into two categories: *base-load* and *peak-load* plants. Run-of-river stations produce base energy and, if pondage is possible, a relatively low quantity of peak kilowatt-hours too. Storage projects, on the other hand, are mostly peak-load plants, being suitable to contribute, according to the degree of storage, to daily, weekly and seasonal peak production of the power system. As the main objective of creating a reservoir is often, on the contrary, the equalization of the inflows, a consumer needing uniform base load can also be fully satisfied (e.g. for some kinds of electrochemical, metallurgical industries). Dumped storage is especially established for peak power production. Tidal schemes, wave power generators and depression plants can contribute, without any additional energy conversions, to coverage of base demands only.

Base energy is evidently of lower value than peak energy. Power production, due to the intricate economic evaluation of the various kinds of producible kilowatt-hours, uses a cluster of energy types supplying the network.

4 LOW-HEAD RIVER DEVELOPMENTS

4.1 River Channel Projects

4.1.1 Types of General Layout

The simplest type of low-head river plant consists of two parts: the weir (spillway) and the powerhouse. Depending on the physical conditions and on the selected type of structure, the powerhouse may be located within the original channel or in a bay-like enlargement. The shaping of the bay influences the head losses and the uniformity of flow distribution to the turbines.[26] The river barrage on a navigable waterway also incorporates a single or double ship lock, either adjoining the weir or the powerhouse or situated in a by-pass canal (Fig. 5). According to local needs the project may also comprise a boat-lock, a fish-pass, a log-chute, a sediment flushing sluice, further intakes for diverse water requirements and control structures. In some instances, mainly to simplify construction, the barrage is placed into a cut of the river bend (Fig. 6). To ensure the safe release of floods, gated weirs are applied almost exclusively on larger watercourses.

It used to be very common to build an intake structure in front of the powerhouse at the entrance of the headwater bay, consisting of sill, coarse rack, skimmer wall and service bridge (Fig. 7). This arrangement has practically been abandoned and is now adopted only in exceptional cases.

Typical low-head stations fall into four main categories according to the location and the general layout of the powerhouse:

1. Unit-block and twin arrangement
2. Pier-head power station
3. Submersible power plant
4. Underground location

4.1.2 Unit-Block and Twin Arrangement

The machine sets are installed into an open-air powerhouse block, or divided between powerhouses adjoining the weir on either side (Fig. 8).

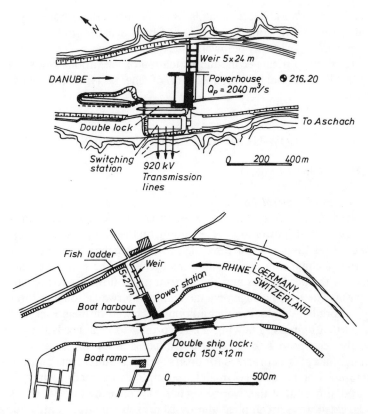

FIG. 5.   River barrage on navigable waterway. Upper drawing: ship lock directly joins the other structures; Aschach, Danube, Austria, 1964, 2040 m³/s, 15 m, 286 MW.[26a,b] Lower drawing: ship lock in by-pass canal; Birsfelden, High Rhine, Switzerland, 1954, 1500 m³/s, 4·9–9·2 m, 85 MW.[26c]

The latter, the so-called twin arrangement, was, in a few cases, chosen when the river at the power site constituted the frontier between two countries and, although the project was a joint undertaking, both wanted to possess a separate powerhouse (Fig. 9).[27] Furthermore, it is sometimes useful to increase the flood-discharging capacity of the weir by locating it into the main current of the river (Fig. 10)[28,28a,28b]

In the conventional solution the block-type power station is equipped with conventional machine sets, with regular setting, i.e. the roof of the spiral case lies lower than the lowest headwater level. The conventional aggregate comprises a vertical-shaft axial (Kaplan or propeller) turbine

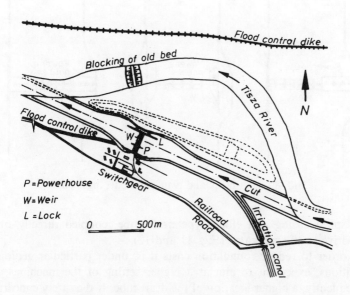

FIG. 6. River barrage located in a cut of the river channel; Tiszalök, Tisza, Hungary, 1953, 300 m³/s, 0–7·5 m, 12 MW.

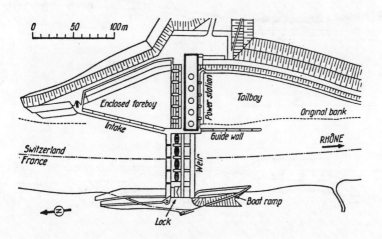

FIG. 7. Power plant with enclosed headwater bay; Chancy-Pougny, Rhône, France.

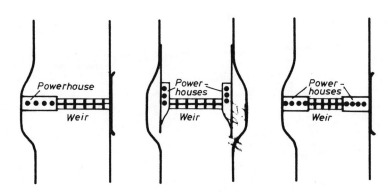

FIG. 8.   Unit-block and twin-station arrangements.

with spiral casing and elbow-type draft tube, coupled directly or by gear-drive to the generator (Figs 11 and 12).

In order to reduce foundation costs it is, under particular geological conditions, expedient to aim at a higher setting of the machines and, consequently, a higher location of the draft tube. If the safety concerning

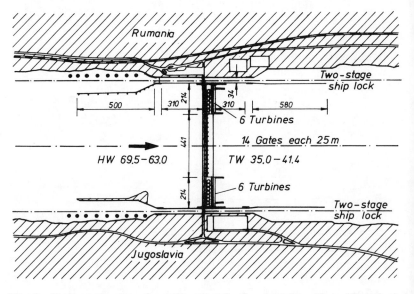

FIG. 9.   Twin station on a boundary stretch of a river; Iron Gate, Yugoslavia/Rumania, Danube, 8700 m³/s, 27·2 m, 2050 MW.[27]

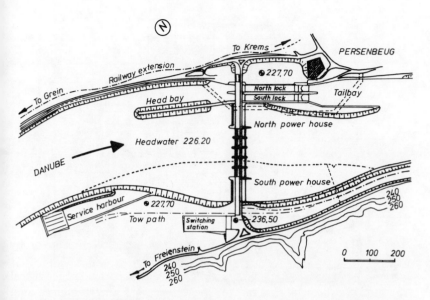

FIG. 10. Twin arrangement chosen for hydraulic reasons; Ybbs-Persenbeug, Danube, Austria, 1959, 2160 m³/s, 10·6 (6–14) m, 200 MW.[28]

cavitation is not jeopardized (see Section 7), even a so-called siphon setting[29,30] may prove to be an optimal solution with the spiral case roof elevated above the minimum headwater level (Figs 13 and 14).[31]

A remarkable solution for reducing construction costs has been achieved by the construction of spillway conduits within the powerhouse proper, thus rendering possible a reduction in the width of the weir or even its complete omission. The spillway conduits releasing partly or completely the excess flow $Q - Q_p$ may serve two more purposes: firstly, the high-velocity jets, released through the conduits and submerged into the tailwater, depress the tailwater level at the draft tube exits and consequently increase the effective turbine head; secondly, in the case of a sudden load rejection, by using automatic gate control devices synchronized with the turbine governors, the generation of surges[32,33] (positive in headwater and negative in tailwater) can be avoided or at least significantly damped. This beneficial effect is of paramount importance on navigable watercourses. The main types of arrangement are exemplified here by a powerhouse with elevated[34] (Fig. 15) and deep-located[35] (Fig. 16) spillway conduits. With the diversion-type of development it is possible for the capacity of the conduits to equal the plant design flow Q_p, so that construction of a weir

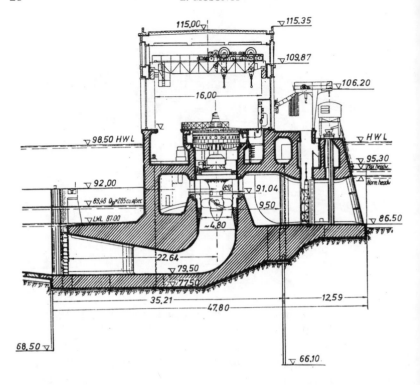

FIG. 11. Cross section of a typical low-head powerhouse equipped with conventional Kaplan machine sets; Tiszalök plant (see Fig. 6).

or a spillway at the station is not necessary at all. Outstanding examples for similar solutions are power plants on the navigable Grand Canal d'Alsace along the upper Rhine.

The head-increasing effect of the overflow jet has been overestimated in the past as it is restricted to periods when $Q > Q_p$ and is induced only by that portion of the excess flow which spills directly in front of the draft tubes (see also Section 4.1.4). The flow through the adjoining weir, especially in the case of a long dividing wall, has no or only a very slight bearing on the tailwater level at the draft tube exits.

In the last decades the bulb-type tubular aggregates have become the favoured machines in block powerhouses for harnessing heads in the lower portion of the low-head range, i.e. up to about 25 m (Fig. 17).[36] The tubular turbine adapted originally, in its first version (with a rim-generator), to the submersible powerhouse design has undergone a multi-phase

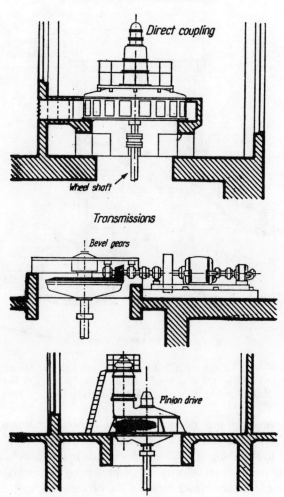

FIG. 12. Coupling of turbine to generator.

evolution during the last five decades. Some technological problems have retarded its wider application so that the more common tubular set at present is the bulb-type turbine. But a very promising revival of the rim-generator turbines has recently been witnessed, i.e. its improved version, the so-called STRAFLO turbine operating well for heads of up to 15–20 m (Fig. 18).[37] Finally, it has to be mentioned that the S-type tubular turbine (coupled to an open-air generator) is also favoured in the lower portion of

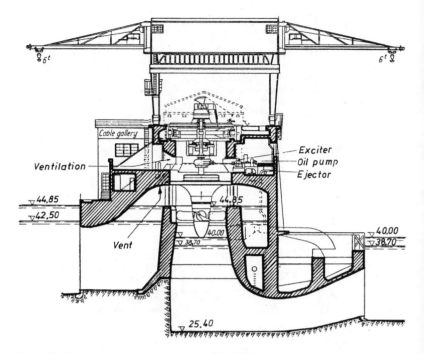

FIG. 13. Low-head station equipped with conventional Kaplan units in elevated (siphon) setting; Vargön, Göta-Älv, Sweden, 1934, 700 m³/s, 3·3–4·3 m, 26 MW. (After a publication of the KMW turbine factory, 1946.)

the low-head range and designed, almost without exception,[37a] for low-capacity units (Fig. 19).

In order to avoid disturbing whirl formation in front of the powerhouse and to obtain a uniform velocity inflow pattern to the turbines, the proper shaping of the dividing wall between the powerhouse and the adjoining structure(s)—weir, ship lock, etc.—is of great importance.[38,38a,38b,53]

4.1.3 Pier-Head Layout

The tendency to reduce the enlargement of the river bed required for the powerhouse bay resulted, about 50 years ago, in a combined design of weir and powerhouse when the machines were placed in the piers of the weir (Figs 20 and 21), each pier housing one or two generating units. The width of a pier accommodating generating machinery is, of course, greater than that of a pier of a normal gated weir; nevertheless, the total length of the plant is reduced. The designers also argued that the layout provided for a

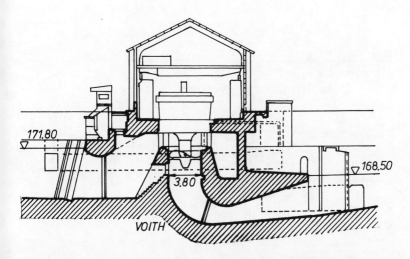

FIG. 14. Station with conventional Kaplan aggregate in siphon setting for utilizing very low head; Randersacker, Main, German Federal Republic, 100 m³/s, 2·7–3·3 m, 2 MW.[31]

less disturbed inflow to the turbines when compared with the curved flow pattern in front of an in-bay located block powerhouse.[39–42]

Despite the suggested beneficial properties, this design has only been used on two small rivers in Europe (Drau and Inn). Lively discussion had been, for a long time, taking place about the advantages and drawbacks of this solution.[28,28a,43,44] It is plausible that, with growing unit capacities, both advantages, i.e. saving in construction width and higher hydraulic inflow efficiency, are diminishing or even disappearing. Also, the designers of equipment (mechanical and electrical) and, even more, the engineers responsible for the operation, are for various reasons averse to a scattered powerhouse. Yet this solution can still be justified in exceptional conditions: no possibility for any widening of the river bed, small or medium-size plant with fairly small units, and where piers are not seriously endangered by drifting ice.

4.1.4 Submersible Power Station

The powerhouse incorporated in the body of the overflow weir is the characteristic feature of this type of plant. Passage of floods can be alleviated by additional bottom outlets (Fig. 22).[45] In order to increase the spilling capacity of the weir and to prevent an excessive rise in backwater

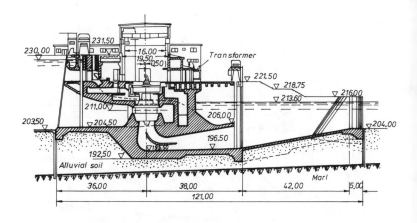

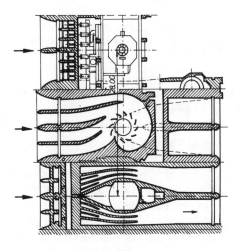

Fig. 15. Block power station with conventional Kaplan aggregates and elevated spillway conduits; Ottmarsheim, Upper Rhine–Grand Canal d'Alsace, France, 1952, 1400 m³/s, 15·5 m (max.), 144 MW.[34]

levels, the weir crest is usually well below the designed headwater elevation. Hence regulation by crest control gates is indispensable. If bottom outlets are also necessary, they alternate with the generating units. In the first stations of this type rim-generator tubular turbines were installed, as shown in Fig. 22, while later bulb-type sets were given priority (Fig. 23). If the attainable head makes the construction of a sufficiently high dam

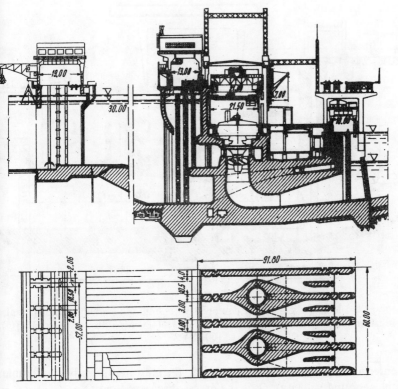

FIG. 16. Block power station with conventional Kaplan units and deep-intake
release (spillway) flumes; Volgograd, Volga, USSR, 19 m, 2530 MW.[35]

possible, vertical-axis (conventional) aggregates can also be chosen (Fig.
24).[46]

The decision about the use of a submersible layout depends primarily
on the discharge and sediment and ice regimes.[47,48] Sealing problems,
considered earlier as possible drawbacks of this arrangement, no longer
exist. During periods when $Q > Q_p$ the head-increasing effect of the
overflowing jet can be taken into account. The formula derived by the
author gives, with fair approximation, the tailwater depression:

$$z \simeq \frac{v}{g}\left(v_e \frac{Q_e}{Q} - v\right) \text{ [m]} \tag{24a}$$

where v is the mean velocity in the tailwater (after mixing), v_e is the

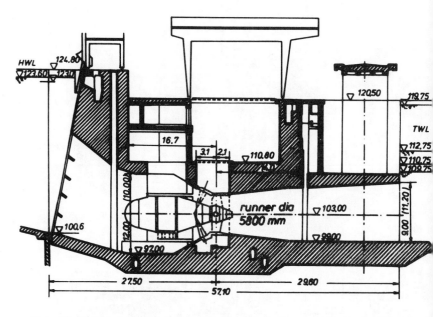

FIG. 17. Block power station equipped with bulb-type tubular Kaplan units; Iffezheim, Upper Rhine, France/FRG, 1977, 1100 m³/s, 10·8 m (max. 12·5 m), 100 MW.[36]

entering velocity of the excess flow into the tailwater (slightly smaller than the spouting velocity), Q_e is the excess discharge, and $Q = Q_p + Q_e =$ total flow.

4.1.5 Underground Layout

The development of underground hydropower plants shows that, despite a few cases of clearly defence objectives, the main intention was to make use of technological and economic advantages. Though the beneficial effects of an underground location are fairly obvious in the majority of high-head plants, it has been found that, under specific topographical and geological conditions, this layout is favourable also for low-head stations (Fig. 25).

The terrain along the river banks and sharp bends which can be cut by short tunnels is in some instances favourable for an underground location of the plant. When comparing an open-air with an underground variant it may be found that for economic reasons, i.e. to reduce investment and operating costs, the latter has to be given priority. Moreover, even when the economic balance is not in favour of the underground alternative, for

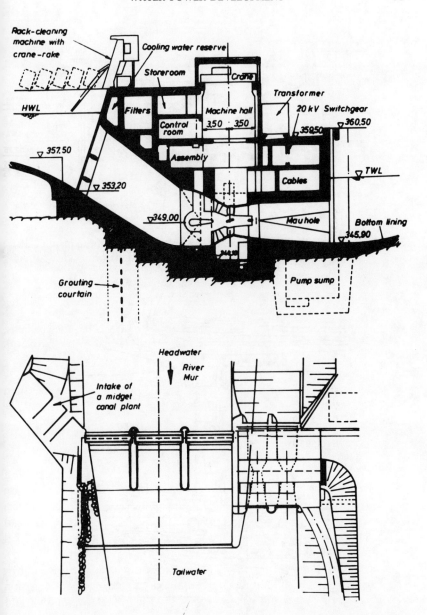

FIG. 18. Block power station with rim-generator Kaplan (STRAFLO) aggregates; Weinzödl, Mur, Austria, 1982, 180 m^3/s, 9·8 m, 16 MW.[37]

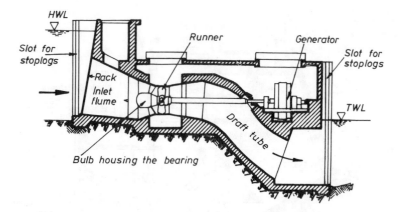

FIG. 19. Low-capacity S-type (propeller or Kaplan) set in low-head plant (TUBE turbine of Allis-Chalmers Corporation).

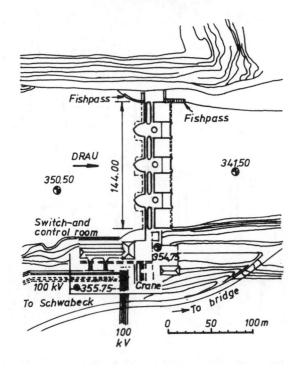

FIG. 20. Layout of pier-head plant; Lawamünd, Drau, Austria, 1944, 380 m^3/s, 9·0 m, 23 MW.[39,40]

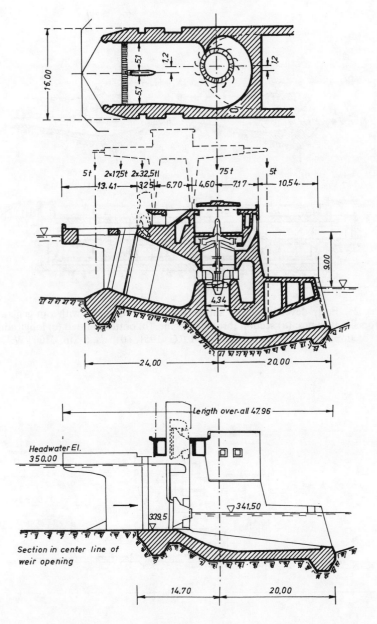

FIG. 21. Lawamünd pier-head station (layout on Fig. 20); horizontal section through the pier, cross-sections of pier and gated weir.[39,40]

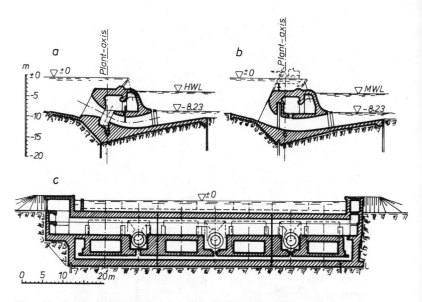

FIG. 22. Early development of submersible plant equipped with rim-generator propeller turbines: (a) section through turbine; (b) bottom outlet; (c) longitudinal section; Rott-Freilassing, Saalach, FRG, 1951, 60 m³/s, 8·5 m, 4100 kW.[45]

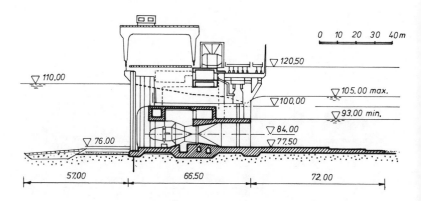

FIG. 23. Submersible station equipped with twenty bulb-type Kaplan sets; Kiev, Dnepr, USSR, 9·4 m, 320 MW. (After E. Gruner.)

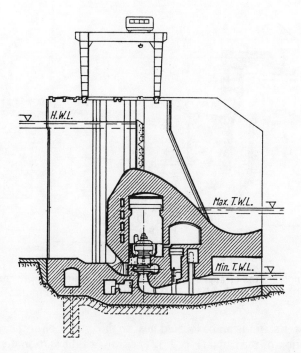

FIG. 24. Submersible station with conventional Kaplan units located within the weir body; Pavlovsk, USSR.[46]

the sake of environmental and landscape protection this solution may still be adopted.

Figure 26 shows a very-low-head project where bulb-type tubular aggregates are installed.

4.2 Diversion Type Projects

4.2.1 Power Canal
Physical (hydrological, topographical and geological) conditions, environmental and economic requirements, further anthropogenic developments, even political aspects, may favour this solution. The flow is partly (seldom completely) blocked and, if necessary, impounded in the watercourse by a diversion weir and is diverted into a power canal which, at a shorter or longer distance, again joins the river. If the diversion is short it is a loop development or power branch of the project.

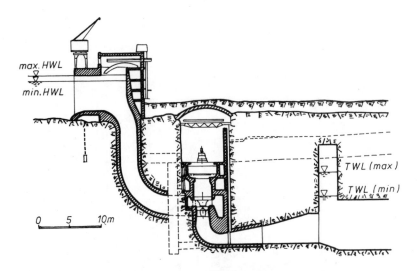

FIG. 25. Underground arrangement of low-head plant equipped with conventional units. (After K. I. P. Wittrock and K. G. G. Pira.)

The power station can be located next to the intake, or at the outlet, or at any section of the canal (Fig. 27). The siting of the station depends on the complex effects of the above listed factors, where control of the groundwater table may have a predominating role.

In preventing, or rather reducing, intrusion of bedload into the canal,

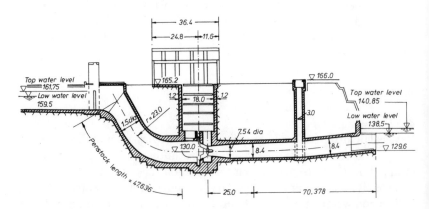

FIG. 26. Low-head underground station housing a single tubular bulb-type aggregate; Shingo 2, Agano, Japan, 200 m³/s, ~23 m, ~41 MW.[49]

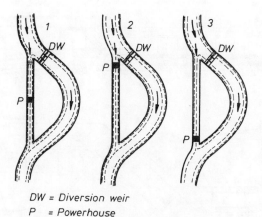

DW = Diversion weir
P = Powerhouse

Fig. 27. Location of the station in the power canal.

the proper choice of intake site is of paramount importance. To deal with this problem, especially on rivers with heavy sediment burden, careful analyses and often scale model tests are indispensable.[50–53] Various structural designs have been proposed and implemented to prevent intolerable bedload intrusion into the canal.[51,52,54]

If physical conditions demand it, protection measures against entrance of ice and floating debris also have to be taken.

Thus an intake structure (canal headwork) may consist of: entrance sill of 0·5–3·0 m height (with or without sediment flushing conduits); skimmer wall submerging to a depth of 0·5–1·0 m below deepest water level, to retain material floating on the water surface; coarse rack for the detention of slush ice and other underwater waste; intake gate to control inflow into the canal (in most cases, however, not required). Figure 28 shows headworks with a sediment flushing system and intake gates. Figure 29 shows a simpler design when physical conditions do not require a sophisticated sediment-control arrangement. Whenever possible, the rack or even an intake structure may be completely dispensed with for economic reasons.

The head loss at the intake consists of entrance loss and rack loss (if any). Assessing the inflow velocity v between 0·8 and 1·2 m/s, and using the author's estimate, at a properly shaped entrance, under usual conditions, a head drop of 2–10 cm can be assumed.[55] Several formulae are available for calculation of the rack loss.[56–58] For example, after Kirschmer:

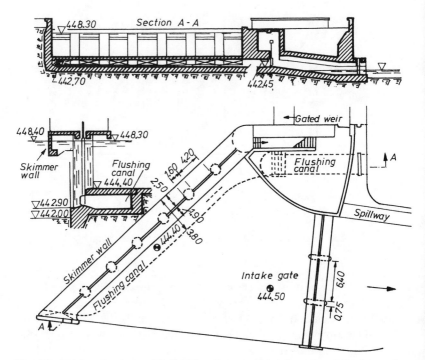

FIG. 28. Intake structure with flushing flume and intake gates; Mixnitz project, Mur river, Austria.

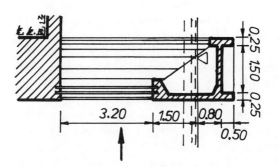

FIG. 29. Simple intake structure: sill, rack and skimmer wall.

$$\Delta h_r = \phi \left(\frac{s}{b}\right)^{4/3} \sin \alpha \left(\frac{v^2}{2g}\right) \text{ [m]} \qquad (25)$$

where Fig. 30 explains the bar characteristics. (Inflow velocity v is the mean velocity in the rack section without the rack.)

The design of the power canal aims to reduce the head losses to a minimum. Therefore power canals are mostly lined. If the canal is not lined stable-regime conditions have to be ensured, i.e. neither sediment deposit nor erosion should occur. The permissible minimum and maximum velocities can be obtained from the literature.[60-62] The common rack loss equations give reliable values only when the inflow is parallel to the longitudinal centre line of the rack bar cross-section, while in the case of oblique inflow, with growing inflow angles and increasing bar lengths, the rack loss rapidly increases.[59] This is why hydraulic conditions providing for parallel inflow have to be ensured as far as possible. This is of even greater importance in the case of powerhouse racks (see Section 9).

The following guidelines have to be considered in the designing and dimensioning of the canal:

— Head losses have to be minimized as far as possible, yet within the economically tolerable limit. Hence, the flow velocity, generally, does not surpass 1·0–1·5 m/s. Thus, in most cases, a lining of the canal is desired (stone, asphalt, concrete, reinforced concrete). It also has to be borne in mind that very high velocities inducing high turbulence and intensive vortices may, in cold climates, alleviate the formation of slush ice clogging the powerhouse rack.

— The flow velocity should not be lower than the critical value for silt deposition. The permissible lowest velocity is usually between 0·3

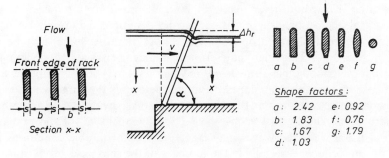

FIG. 30. Definition sketch for Kirschmer's rack loss equation.

and 0·7 m/s. Earth power canals have to be designed as stable-sediment-regime channels.

— Backwater elevations and surge waves generated by sudden load rejection in the powerhouse must not rise above the canal banks (in cuts) and the dykes (in fills).

— Canal flow must not detrimentally influence the groundwater table. Seepage control measures (lining, sheet piling, drainage system, groundwater recharging canals and wells) can be considered to balance the unfavourable effects of the diversion.[63,64] Drainage and recharge systems have been applied at the Rhône Development (Fig. 31).

— Last but not least, the environmental impacts of the canal have to be evaluated and, if necessary, protective measures have to be envisaged (forestation or reforestation of the canal banks and land behind the dykes; stone revetments or loose rip-rap paving of the canal slopes—thus even tolerating a higher head loss—in order to permit the formation of small biotopes).

Additional design criteria have to be considered for navigable power canals. The main prerequisites for safe navigation are: flow velocity adjusted to the envisaged type of vessels, protection of banks and embankments against waves caused by the vessels, measures to avoid or reduce surges which otherwise come into being in cases of so-called load rejection.

This latter phenomenon and the possible control measures (mentioned already in Section 4.1.2) deserve further elucidation. Maximum positive surges arise in the headrace and negative surge waves are generated in the tailrace of a power canal when an emergency closure occurs, i.e. in the case of a network failure or of some operational breakdown in the powerhouse. Similar surges may also occur in the headwater reach and in the tailwater section of a run-of-river plant.[32,33] Without providing for special surge control, the positive surge wave at low-head stations may reach a height of 1–1·5 m, while the drop in tailwater level (negative surge) may be as much as 0·8–1·2 m which may seriously damage banks and slope linings. High waves and quick drawdowns in water level may tear up unprotected banks or embankment slopes. In navigable canals the surge waves may endanger vessels, pushing them against the banks, dividing walls and other structures, and against each other. The superposition[65] of the primary waves and reflected waves may create very unfavourable conditions in certain sections, especially in the forebays of the ship locks, even causing serious collisions. Navigation can seriously be impaired when the lock is

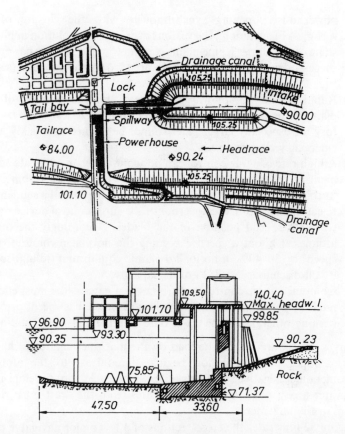

FIG. 31. Bauchastel plant, Rhône, France, 1963, six conventional Kaplan units, spillway with four top leaf gates and four bottom outlets, 2100 m^3/s, 11·4 m, 192 MW. Drainage and recharging canals (and wells) provide for the necessary groundwater equilibrium along the Lower Rhône canalization. (After publications of the Compagnie Nationale du Rhône.)

situated next to the powerhouse. Scale model tests are suitable for clarifying surge patterns and establishing proper control measures.[53,66-68]

Various methods have been attempted at both canal stations and run-of-river plants to damp surges:

(1) Use of water rheostat, manually or automatically controlled. (Obsolete, but still used for test runs and warranty measurements.)

(2) Automatically operating relief orifices or relief conduits within the

powerhouse, opening synchronously with the closure of the turbines. From among the various proposals the solution applied in some of the Upper Rhine plants (Grand Canal d'Alsace) has proved successful where spillway conduits are used for this function (see e.g. Fig. 15).

(3) Rapid opening of spillway gates (or of the weir of a run-of-river plant). This measure alone, however, in most cases is not satisfactory for especially larger gates cannot be opened as quickly as the emergency closure occurs.

(4) A high rate of surge control can be achieved by an adequate adjustment of the guide and runner blades of the Kaplan turbine. This 'turbine surge control' works as follows. In the case of a sudden load rejection the automatic governor of the surge control device regulates for a special position of the blades. Accordingly the turbine attains, at about a speed exceeding the normal revolution (rated speed) by 30–40%, a no-load dynamic equilibrium (idling) so that it still discharges 50–75% of its capacity.

(5) Combined surge control by turbines and gates is the most effective method for damping surges. This process has been perfected (among others) at some up-to-date Danube and Mosel stations.

Surges are induced when the load on the machines suddenly changes. Thus if rapid peak operation begins, in contrast to the aforesaid phenomenon, a negative surge wave (i.e. water level drop) develops in the headwater and a positive surge propagates from the draft ports downstream in the tailwater.[33]

Figure 32 illustrates the cross-sections of a few major navigable power canals.

4.2.2 Power Station

The power station of a diversion type project comprises two main structures: powerhouse and spillway. The latter provides for discharging of the excess flow, or the complete power flow, during periods when the generating sets are partly or entirely out of operation. Thus, since the floods are released by the diversion weir (in the river), the station spillway has to be dimensioned—similarly to the canal—for the plant design flow Q_p. The structure is usually a gated weir which permits the maintenance of a constant headwater level or its regulation within the prescribed limits. In the case of small plants, less expensive overflow weirs are sometimes also constructed. Reduction in the spillway width or an abandonment of this

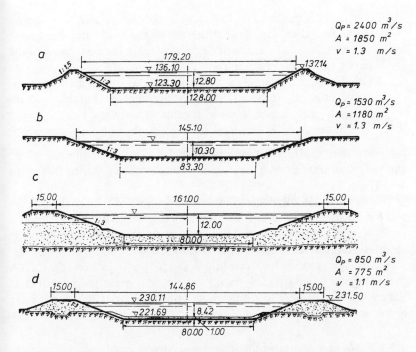

FIG. 32. Standard cross-sections of some major navigable canals: (a) Czecho-slovak–Hungarian Danube project (under construction); (b) Donzère-Mondragon diversion, Lower Rhône, France; (c) tailwater stretch of the Kembs plant; (d) headrace of the Ottmarsheim plant, Grand Canal d'Alsace, Upper Rhine development.

structure can be achieved by establishing a powerhouse with release conduits (Figs 15 and 16), or by choosing a submersible solution (Figs 22–24).

To reduce width and costs of the spillway, all types of axial turbines can be designed for discharging a high rate of flow even when no power generation takes place (see Section 7.3).

Under certain circumstances, however, there is no need at all to provide for a spillway or release conduits. The necessity of discharging the full or partial power flow is justified in the following cases:

(a) The canal accommodates two or more stations, and hence the operation of the available powerhouse(s) must not be jeopardized by the station being out of service.

(b) The canal is bordered by low-crest banks or embankments sloping in the flow direction, so that the rise of the headwater due to the decrease or discontinuation of flow may result in overtopping of the banks and dykes respectively.

(c) Other requirements may account for a need for permanent fresh-water flow (e.g. fisheries, groundwater control, potable and industrial water demands, environmental quality, etc.).

(d) The power canal is also used for navigation.

It follows that, if a single station in a canal is close to the diversion weir and no constraints listed above exist, a canal spillway can be dispensed with (Fig. 33).

The station is usually located in an enlargement of the canal, with the spillway (if present) next to it or even dividing the power plant into two parts. A small project, in several instances, comprises a side spillway operating as overflow crest weir and, necessarily, a bottom outlet (Fig. 34).

According to specific local conditions and project objectives, the station can be supplemented with fish-pass, ice chute, boat sluice and/or ship lock. Recently, even the very low differences of head in irrigation canals have begun to be harnessed.

4.2.3 Protection Measures for Abandoned Stream bed

When calculating the achievable power and energy, with special regard to the minimum available (firm) power, the duty flow in the abandoned river bed has to be properly estimated, for this quantity has to be deducted from the minimum river flow to obtain the minimum power flow. The duty flow (also called compensation flow) depends on water demands along the by-passed river stretch and on environmental requirements.

It may be desirable to establish sills and/or control weirs in the abandoned river bed in order to maintain water levels higher than those pertaining to the duty flow. Instructive examples for similar measures are demonstrated by the Upper Rhine Development.[69]

It has to be borne in mind that the hydrological regime of the abandoned river stretch, especially if a long one, undergoes a complete change: its hydrograph gains a highly flashy character requiring a careful study of the passage of floods and the consequences of the altered sediment transport.

4.3 Preliminary Studies and Planning Criteria

Selection of the power development site necessitates detailed topographical surveying, thorough geological investigations and evaluation of the fluvial

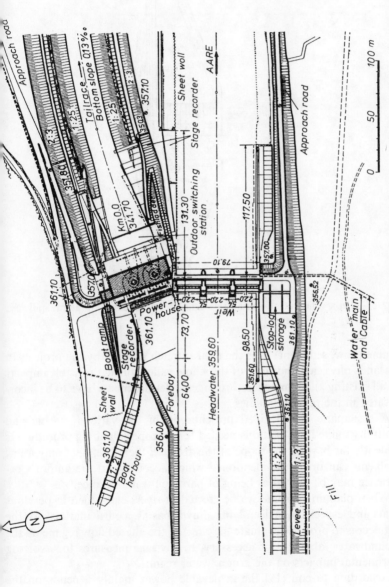

FIG. 33. Canal plant linked to the diversion weir, Rupperswil-Auenstein, Aare, Switzerland, 350 m³/s, 11 m, 33.7 MW. (After a publication of Kraftwerk Rupperswil-Auenstein AG, 1949.)

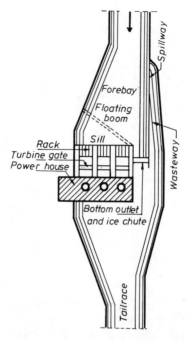

FIG. 34. Typical layout of a small-size canal development with lateral spillway.

regime (flow, sediment) over the entire stretch affected by the project. In lowlands physical conditions of the whole valley, further possible impacts on all existing and planned anthropogenic developments have to be incorporated in the relevant studies.

A systematic and detailed presentation of the various specific and multi-discipline investigations needed for a proper planning procedure is evidently far beyond the scope of this treatise; therefore, here only a few problems can briefly be mentioned which according to the author's experience may be judged as being of paramount importance.

When planning schemes to be erected in rivers embedded in pervious strata and, especially, in deep-lying alluvial basins, a particular task for the engineer is to properly evaluate the effect of the altered flow regime on the groundwater table and, if necessary, to envisage measures for avoiding detrimental impacts on the groundwater regime.

According to eqn (19) the achievable power mainly depends on the magnitude of the two principal parameters, i.e. on the available flow and head, while the factor c expresses the complex resultant of losses. The analysis of the *flow regime* has to result in discharge duration curves clustered, if necessary, according to type of year and to seasons. It is recommendable to acquire flow hydrographs for long series of years. Since both seasonal and multi-annual variation of flow strongly influences the economic evaluation of the producible energy, if long-term hydrological data are not available, it is at least desirable to plot the flow duration curves for three characteristic years, i.e. for an average, a wet and a dry year. Sometimes the first approach can be made on the basis of a hydrologically typical mean duration curve. In most cases, however, mean values are not acceptable and may even be misleading.

Special attention has to be given to a reliable determination of *floods* and *low flows*. Flood analysis furnishes the basis for the selection of the design flood decisive for dimensioning of the spillway and gated weir whilst a proper estimate for the low discharges is essential for assessing the firm (guaranteed) power. A 95% duration discharge is generally adopted for evaluation of the firm power.

Since, at low-head stations, the head considerably varies, the selection of the design head for the turbines is often an intricate procedure (as opposed to high-head plants) because reasonable efficiency and output have to be ensured for both of the extreme heads.

The principal technological rule that a given natural resource can be exploited, or a predetermined quantity of goods produced, up to a certain limit, more economically by the application of fewer high-capacity than of more low-capacity producers, applies also in hydropower engineering. Consequently, in order to achieve an economic optimum, the natural drop of a selected river stretch has to be utilized by a minimum number of steps, i.e. with the possible maximum head at each plant. Nevertheless, the above expression 'up to a certain limit' means that hydrological, topographical, geological and anthropogenic conditions and, last but not least, environmental constraints dictate a limit for the maximum permissible headwater evaluation. In some cases, given sufficient information, the experienced engineer is able to make a sound decision. If not, variants for various heads have to be elaborated and their technological and economic feasibility and environmental impacts compared.

Foundation and construction problems have to be studied at a very early stage of the planning to avoid technological difficulties and, consequently, unexpected excess costs which may completely negate the results of the feasibility study. Soil mechanical analysis furnishes data concerning the most suitable foundation for the structure method (dry pit construction, groundwater lowering, caisson, pile foundation impervious curtain or sheet pile closure, etc.).

In alluvial rivers, downstream of a single barrage, or the last plant of a cascade, the possibility of intensive degradation of the riverbed must be reckoned with. The profile and progress of degradation can be calculated if sufficient knowledge on the sediment regime is available. If degradation is too strong, or no other protective measures are acceptable, an artificial supply of sediment may prove successful to re-establish the natural sediment transport equilibrium.[70]

It is important to provide for a safe release of the floods occurring during the construction period. The construction method and schedule should therefore be based, besides detailed hydrological and hydraulic investigations and possibly scale model tests, also on nautical studies, if the plant is constructed on a navigable river.

In stream or canal developments incorporating navigation locks the general arrangement of the plant, the shaping of the lock forebays and the length of the separating piers or walls have to be designed so that no disturbing cross currents affecting the vessels occur. This holds good for the hydraulic dimensioning and shaping of the filling and emptying systems of the locks proper. The relevant guidelines in the Federal Republic of Germany prescribe—if specific investigations and model tests do not advocate other limitations—that the velocity of the cross current along the path of the vessels should not exceed 0.3 m/s. In most cases, model tests are indispensable.[53]

5 MAGNITUDE AND EVALUATION OF PRODUCIBLE POWER AND ENERGY

5.1 Plant Potential

5.1.1 Main Plant Parameters
To every combination of headwater level (HWL) and tailwater level (TWL) pertains a static head H_{st} upon the turbine proper and a net head H which is the effective head available for the turbine for power conversion. Thus, as Fig. 35 illustrates:

$$H = H_0 + v_0^2/2g - \Sigma\Delta h - v^2/2g \qquad (26)$$

where the intake loss $\Sigma\Delta h$ is the sum of the entrance loss Δh_e, rack loss Δh_r, slot loss Δh_s and inlet flume friction loss Δh_f; v_0 is the inflow velocity and

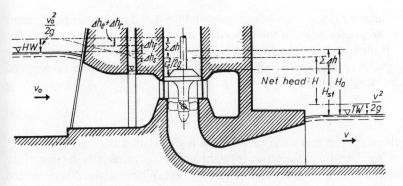

FIG. 35. Definition sketch of the net head.

v is the exit velocity from the draft tube. The losses in the spiral case (with tubular machine in the bulb) and in the draft tube are included in the turbine efficiency because the spiral case (bulb flume) and draft tube are, in spite of their civil engineering character, hydromechanically and operationally inseparable parts of the turbine. Accordingly the proper internal shaping of these two structures guaranteeing efficiencies and the necessary operational safety margins are designed by the turbine supplier.

Since at low-head stations the head significantly varies, a certain magnitude within the head range is selected by the designer of the turbine as rated head; it is the lowest head at which the turbine with full-gate discharge can produce the power (turbine rated power) providing the rated capacity of the generator. Maximum efficiency does not necessarily pertain to the rated head (sometimes also called design head). The term critical head is used for two very different operational cases, i.e. for the overload limit (US Department of the Interior), or for the extreme levels—highest HWL at lowest TWL—indicating sometimes cavitational limit. Composite references on definitions are available.[71,72]

Rated discharge is the full-gate discharge at the rated head (also termed design discharge). In most cases, i.e. when the generating set is not designed for overload, the rated discharge is referred to as turbine discharge capacity.

All these definitions and terms pertain to the operational (rated) speed n which is determined by the network frequency and generator design:

$$n = \frac{60f}{p_\mathrm{p}} \ [\mathrm{rpm}] \tag{27}$$

where n is obtained in revolutions per minute (rpm) when the network frequency f is given in cycles per second (cps) and p_p denotes the number of the pair of poles of the generator.

If the turbine is not directly coupled to the generator, i.e. a gear drive steps up turbine speed to generator speed n_g, a lower turbine rated speed can be selected:

$$n = \frac{n_g}{v} = \frac{60f}{p_p} \text{ [rpm]} \qquad (28)$$

where v is the gear ratio (usually not higher than 8 to 10).

The plant discharge capacity (plant design flow) is usually the sum of the rated discharges of all turbines of the plant. When overload is permitted a higher discharge capacity (maximum discharge) is also allocated to the plant. The installed (rated) plant capacity and, in the case of overload, the maximum plant capacity can accordingly be defined.

5.1.2 Estimate of Overall Plant Efficiency

Already in the preliminary stage of the planning procedure it is of great importance to make a reliable estimate of the producible power and energy. For this, the overall plant efficiency has first to be assessed.

From the project gross head all losses both in the headwater and tailwater reaches (with diversion schemes also the canal losses) have to be subtracted to obtain the powerhouse gross head (H_0 in Fig. 35); evidently no estimates of general validity can be given for this. When, however, based on individual project features, the H_0 value is at least roughly determined, the overall plant efficiency can approximately be evaluated as follows:

(1) The rack loss, when excluding oblique inflow due to deficient design, varies usually between 5 and 20 cm. The slot and flume friction losses, mostly not exceeding 2–5 cm, can be neglected in the preliminary design. To remain on the safe side the intake kinetic energy ($v_0^2/2g$) should be neglected and the exit kinetic energy ($v^2/2g$) can be valued at 10–20 cm. This, according to eqn (26), permits a first assessment of the net heads pertaining to the varying gross heads.

(2) The highest operational efficiency of an up-to-date turbine, depending on its output, can be estimated between 0·85 and 0·92.

(3) The full-load efficiency of a three-phase synchronous generator η_G, depending on its capacity and on the network power factor, varies between 0·92 and 0·97.

(4) The transformer efficiency lies within the range 0·94–0·98.

Thus, in the planning phase the plant power output can be calculated, with reference to eqn (19):

$$P = 9\cdot81\,\eta_T\eta_G\eta_{tr}QH = 9\cdot81\,\eta QH = c_0QH\,[\text{kW}] \qquad (29)$$

where H (m) denotes the net head upon the turbine, Q (m^3/s) the available and utilizable discharge at any arbitrary time, η the overall plant efficiency and c_0 the overall power coefficient of the plant. For plants to be equipped with medium or large units (exceeding 5 MW) a c_0 coefficient between 8·0 and 8·6 can be anticipated. In the field of small hydro, down to mini and micro projects, the coefficient may drop from 8·0 to 7·2. Extreme values beyond the above limits also occur. It is recommended to assess conservative values in the feasibility studies and all preliminary phases of planning. For the final design the manufacturers furnish the efficiency data.

The installed capacity in hydropower stations usually equals the rated capacity as against other types of power plants which usually also incorporate stand-by generating aggregates.

5.1.3 Power Curve and Energy Production Diagram

It is usual to display power versus time derived from the discharge duration curve of an average year or from a fictitious duration diagram established from a series of annual hydrographs. In the case of diversion projects the discharge duration curve has to be reduced according to a constant or varying duty flow demand, if any. Figure 36 shows the construction of the power curve. On the horizontal axis one year is shown; on the vertical axis both discharges and heads are indicated. With the help of the rating curve adapted to the station site, the duration curve of the water stages is established. The assumption of a perfect run-of-river operation leads to three consequences: (1) the natural stage duration curve becomes the tailwater duration curve after the completion of the plant; (2) the headwater elevation does not change; (3) the difference between the river flow and discharge utilized by the turbines is continuously spilled by the weir or other outlets for excess flow. A pre-estimated value of the $\Sigma\Delta h$ has to be abstracted from the planned headwater level and, consequently, the difference between this lower level and the tailwater duration curve furnishes, according to eqn (26), with fair approximation, the net head available for the turbines. The net head duration curve can be plotted separately.

Since the corresponding values of available discharges and net heads are known, the plant discharge capacity and the turbine design head can be selected. The power pertaining to any discharge $Q < Q_p$ is obtained by

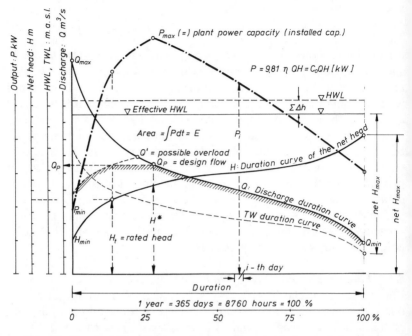

FIG. 36. Construction of the power curve for a typical run-of-river plant.

eqn (29), while during the periods of $Q > Q_p$ the utilizable discharge is obviously restricted. If the rated head is selected as Q_p, the maximum output, i.e. the plant discharge capacity, is

$$P_{max} = c_0 Q_p H_t \text{ [kW]} \tag{30}$$

since, when Q exceeds Q_p, the available head is smaller than H_t, and the turbine can discharge only Q, smaller than Q_p.

At very-low-head plants a rated head lower than the one related to the design flow (H^*) (Fig. 36) is often chosen to avoid unreasonable diminution of the utilizable discharge and producible power at the lowest head. (The value of P_{max} is also affected by the significant drop of efficiency pertaining to lowest heads.) Owing to such a selection the reduction of the utilizable discharges is restricted to the period when H lies between H_t and H_{min}. Accordingly, when the net head lies between H_t and H^*, both turbines and generators can be dimensioned for the hydrologically achievable overload (resulting from Q').

The area bordered by the power curve gives the producible energy in the selected year:

$$E = \int P \, dt \, [\text{kW}] \tag{31}$$

If the daily outputs P_i kW are read from the diagram the annual energy can be computed as

$$E = 24 \sum_{i=1}^{365} P_i [\text{kWh}] \tag{32}$$

The plant design flow (plant discharge capacity) is chosen on the basis of experience (see Section 2.3 and Fig. 2), or evaluated by optimization.[73] For the final design the calculations are usually extended by including suggestions on diverse rated heads, variable machine efficiencies and $\Sigma \Delta h$ losses (function of Q).

If the HWL must be fixed lower than the highest natural flood level at the station site and the floods have to pass through without noticeable impoundment, which is sometimes the condition in lowland rivers,[74] the plant is out of service during the flood periods because no head is created; see Fig. 37(a). On the other hand, if the HWL is much higher than the highest natural flood level (FL), the output fluctuation is smaller and, consequently, the relative magnitude of firm power is higher. Hence, the higher the HWL related to flood elevations, the more valuable is the produced energy; see Fig. 37(b).

The energy production diagram shows the fluctuation of power versus

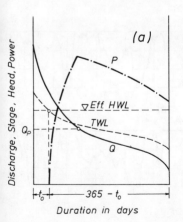
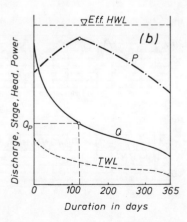

FIG. 37. Power curves: (a) effective HWL below FL; (b) effective HWL much higher than FL.

time in chronological order, and can be plotted either from the stage hydrograph or from the flow hydrograph by allocation of the corresponding power values from the power curve (Fig. 38).

5.2 Plant Load Factor and Annual Utilization Hours

The integrated capacity of the machine, i.e. the installed capacity of the plant, is in most cases, in contrast to thermal stations, selected so that it does not exceed the peak value of the power curve derived from the hydrological design year. Nevertheless, sometimes for one reason or another an increased availability of power is provided for either by the use of more powerful aggregates or by installing reserve (stand-by) units.

An important parameter, the load factor of the plant which is decisive for its economic valuation, can be deduced from the power curve:

$$\alpha_1 = \frac{E}{8760 P_{max}} \left[\frac{kWh/a}{h/a \times kW} = 1 \right] \qquad (33)$$

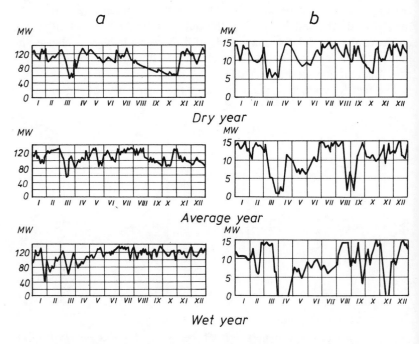

FIG. 38. Energy production diagrams: Eff. HWL (a) above, (b) below FL.

where E denotes the actual energy output per annum and P_{max} is the maximum producible power (mostly equal to the installed capacity). In hydropower engineering the term 'annual utilization hours' is also commonly used to express the same relation. By rearranging eqn (33) we obtain

$$t_u = 8760\,\alpha_l = E/P_{max}\,\text{[h/a]} \qquad (34)$$

indicating the time in hours during which, at maximum load, the same quantity of energy could be produced as the actual annual output (Fig. 39).

Where the installed capacity P_i is higher than the peak load, a further parameter, the plant capacity factor

$$\alpha_c = \frac{E}{8760 P_i}\,\text{[1]} \qquad (35)$$

can be evaluated $(\alpha_c < \alpha_l)$.

6 OPERATION OF CHAIN OF PLANTS

The primary objective of a chain of river barrages located in sequence i.e. forming a cascade, is either to facilitate navigation along a long stretch of river with power production as a secondary objective (river canalization),

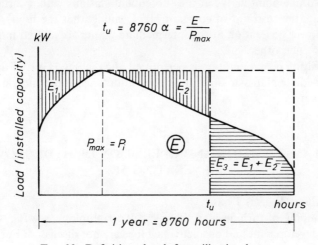

FIG. 39. Definition sketch for utilization hours.

or complete (optimum) harnessing of the hydropower resources of the river stretch in question.

There is a recurrent misconception that the establishment of a chain of power stations on a river navigable in its original state entails only drawbacks for navigation. A theoretical study[75] and *in situ* navigation tests carried out on the Upper Danube[76] proved that creation of backwater ponds may provide noticeable advantages for navigation.

In most cases a flood-routing study is indispensable, since the execution of a river-canalization project may reduce the flood retention capacity of the canalized stretch. Heightening of the levees, provision for retention basins and special operation of the plants provide adequate measures to solve this problem; see the Upper Rhine Development.[77]

If sufficiently large basins can be created at both the first (topmost) and last (lowest) plant of the chain, significant daily and/or weekly peaks can be generated at all intermediate plants as well. The capacity of the storage (first) basin has to be equal to that of the equalizing (last) basin. The latter re-transforms the peak-flow pattern into the natural hydrograph usually required in the river downstream.

The three possible operation modes of a chain of plants are (see Fig. 40):[78,79]

(1) Strictly run-of-river operation (as defined above).
(2) Heaving peak operation is accomplished with constant HW levels at the intermediate plants. Since the peak discharges released by the storage plant arrive at the subsequent stations with a certain time lag, the restraint of constant HWL entails that the hours of peak power generation at the intermediate plants are shifted in relation to each other.
(3) Tilting peak operation is performed at changing HWL of the intermediate plants, in order to achieve simultaneous production of peak energy.

7 CLASSIFICATION AND OPERATION OF HYDROPOWER GENERATING SETS

7.1 Turbine (Wheel) Types
For grouping of turbines and the relevant terminology, see Fig. 41. Historical development and obsolete machines are not discussed. For utilization of low-heads, at the present, mainly axial-inflow wheels are applied.

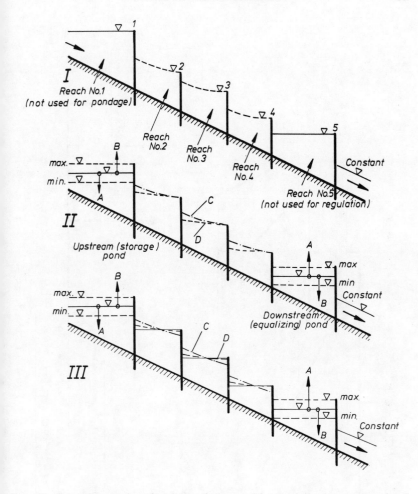

FIG. 40. Operation modes of a chain of plants with two daily storage (pondage) basins, supposing constant stream flow: (I) run-of-river operation; (II) heaving peak operation; (III) tilting peak operation. Backwater profiles in the intermediate reaches: C, during peak operation; D, during off-peak operation. (In case I, in spite of the existence of the basins, no pondage is possible.)

Earlier also radial-inflow turbines were frequently installed in low-head plants, though under certain conditions the use of high-speed Francis machines is still justifiable. Similarly the Dériaz turbine (diagonal-inflow wheel) has been developed for application as a reversible machine (pump-

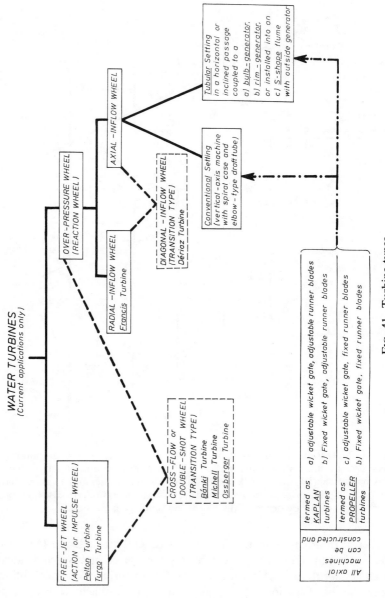

Fig. 41. Turbine types.

turbine), but it is suitable for utilization in the higher part of the low-head range only. Up to the present only a few Dériaz machines have been constructed. In mini and micro stations, besides the above mentioned types, also cross-flow turbines are frequently installed.

7.2 Typical Low-Head Generating Aggregates

The low-head sets can be categorized according to the following main types:

(1) Conventional (i.e. vertical-axis) propeller or Kaplan unit. Inflow from the spiral case into the wicket gate is radial; thereafter the flow is diverted and passes the runner in an axial direction; finally the 'used' water passes through an elbow-type expanding draft tube connecting the wheel with the tailwater. The typical set (conventional Kaplan unit) usually comprises a directly driven vertical-axis generator (Figs 11, 13–16, 21, 24, 25). Figure 42 shows a simplified cross-sectional drawing of a conventional axial wheel and in Fig. 43 the photograph of a large Kaplan runner can be seen. Application head range: up to 50–60 m, exceptionally even higher (e.g. Orlik, Czechoslovakia, $H = 71$ m). The unit capacity exceeds 100 MW at several plants (Ligga, Sweden, 182 MW).

In earlier low-head plants equipped with axial or radial turbines shafts were built instead of spiral cases, and the water was conveyed from the wheel through vertical, diffuser-shaped draft tubes into free-surface tail-water pits. Similar arrangements are still chosen in small projects.

(2) The bulb-generator set comprises an axial (propeller or Kaplan) wheel installed into a horizontal or slightly inclined passage (tube) and a directly coupled or gear-driven generator housed in a steel bulb (bulb unit) (Figs 17, 23, 26). The inflow to the wheel is symmetrical around the bulb and the outflow is effected through the expanding straight draft tube.

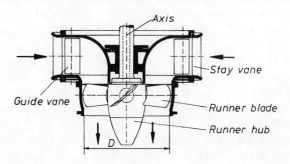

FIG. 42. Cross-section of a conventional axial (Kaplan or propeller) turbine.

FIG. 43. View of a conventional Kaplan runner; mounting in the workshop.
(From issue No. 20, *Voith Forschung und Konstruktion.*)

(During the last decades, as mentioned in Section 4.1.2, the bulb set
penetrated into the lower head field of the conventional unit.) Figure 44
illustrates the cross-section of a bulb unit and Fig. 45 the wicket gate of
a bulb turbine. Unit capacities of 20–40 MW are no longer rare (e.g.
Shingo No. 2 plant, Japan, $H = 22.5$ m, 41 MW; Péage-de-Rousillon,
France, $H = 12.5$ m, 40 MW; Altenwörth, Austria, $H = 14$ m, 40 MW).
In the case of very low heads it may be necessary to insert a gear drive (e.g.
Mosel plants in the Federal Republic of Germany).

(3) In the rim-generator aggregate the poles of the generator rotor are
bolted to a ring which is fixed to the turbine runner and rotates in a slot
of the tubular passage. The generator stator encircles this slot. The delicate
sealing problems have been solved. Its first application (based on an early
patent of L.F. Harza) was restricted to small units operating under very
low heads in submersible plants (Fig. 22). Its improved version, the
so-called STRAFLO machine, is designed for higher heads and for larger
capacities as well, being applicable in any type of plant (see Fig. 18 where
a block powerhouse accommodates two sets, each of 8 MW). Most of the
rim-generator turbines are single-regulated (propeller type). A considerable progress in this field is manifested by the STRAFLO propeller-type
units of 20 MW already in operation in the Annapolis pilot tidal plant

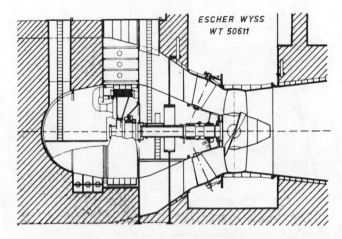

Fig. 44. Cross-section view of a large-size bulb aggregate; Racine plant, Ohio river, USA, $H = 7\,\text{m}$, $P \simeq 25\,\text{MW}$, $D = 7\cdot7\,\text{m}$.[80]

Fig. 45. Assembly of wicket gate of a bulb turbine in the workshop; Parki plant, Sweden, $H = 13\,\text{m}$, $Q = 172\,\text{m}^3/\text{s}$, $P = 20\,\text{MW}$, runner diameter $4\cdot9\,\text{m}$. (After a prospectus of the KWM turbine factory.)

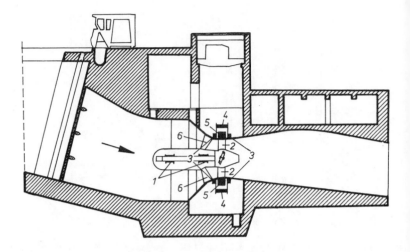

FIG. 46. Sketch of the STRAFLO aggregate: 1, bearings; 2, runner; 3, sealings; 4, generator stator; 5, generator rotor; 6, guide vane. (After a publication of the STRAFLO Group.)

(Nova Scotia, Canada). Figure 46 explains the STRAFLO arrangement. The STRAFLO turbines in the Weinzödl plant (Fig. 18) are double regulated, i.e. they have to be termed STRAFLO-Kaplan machines.

(4) The S-arrangement (see Fig. 19) contains an axial (propeller or Kaplan) wheel accommodated in an S-shaped tubular passage and a generator coupled to it directly or by speed-increasing gear. Vertical-axis S-machines can sometimes be found installed in adequately formed passages. The head range extends from very low heads up to about 15–20 m. Usually the unit capacity does not exceed 5–10 MW. The designs of the Allis-Chalmers Corporation (USA) are termed TUBE turbines. The largest TUBE aggregates are the five approximately 50 MW units of the Ozark plant, Arkansas ($H = 9.8$ m). The S-installation is a preferred solution for mini and micro plants.

(5) The various types of cross-flow turbines (Banki, Michell, Ossberger) are exclusively used in small-size development. As the name indicates, the jet passes the wheel twice (Fig. 47). The cross-flow turbine can efficiently be used for heads from the lowest up to about 200 m, though utilizing low discharges only, so that its capacity is limited to about 1 MW.

(6) Under adequate hydrological and load conditions high-speed vertical-axis Francis machines, even large ones, can still be competitive with axial turbines; however, because of their lower specific speed they are

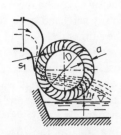 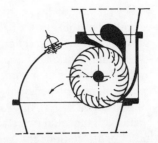

FIG. 47. Cross-flow turbines: (a) schematic drawing of the Banki turbine; (b) cross-section of an Ossberger wheel with vertical inflow.[81-83]

being considered mainly within the upper low-head domain, i.e. at about 40 m or higher (Itaparica, Brazil, $H = 53$ m, three machines each 264 MW; Inga II, Zaire, $H = 63$ m, four 178 MW units; Noxon Rapids, USA, $H = 46$ m, one 128 MW turbine).

7.3 Hydromechanical Principles and Operational Characteristics of Axial Turbines

7.3.1 Basic Principles of Operation

In the conventional set the spiral (scroll) case provides for uniformly distributed inflow along the outer periphery of the wicket gate (guide vane assembly) (Fig. 15) whilst the adjustable guide vanes regulate the discharge (Fig. 42). The scroll case forces the water to move along spiral traces, inducing the formation of an almost ideal irrotational vortex, which promotes the development of a tangential (whirl) component of the flow; the tangential component of the velocity necessary for the proper entrance of the water into the runner is formed along the curved guide vanes.

Water under pressure entering the runner is deviated by its blades, thus exerting impulse forces on them. The rotational components of these blade forces develop a torque revolving the runner and the generator rotor. The pressure part of the entire energy (equal to the net head) is first converted into kinetic energy and then used to a large extent at the exit of the runner. Nevertheless the water leaves the runner with a fairly high velocity. The corresponding kinetic energy would, if no measures were taken, cause an unacceptable loss of the natural resource. Therefore, in order to regain a substantial portion of the wheel exit head, expanding draft tubes are joined to the steel housing (discharge ring) of the runner. Owing to the significant expansion of the draft tube its exit velocity (v in Fig. 35) is much smaller than that from the runner; thus, the difference between the two kinetic heads (reduced by the friction losses within the tube) can be recovered. The draft tube permits the utilization of the entire net head even when the tailwater lies lower than the exit section of the runner (see Figs 13 and 35). The elbow-type configuration of the conventional axial turbines also directs the outflowing water at minimum losses into the river or canal.

The best performance efficiency of the wheel is achieved when, at a selected revolution (rated speed), the water leaving the runner and entering the draft tube has no tangential (whirl) component. It is obvious that only the axial (meridional) component has a bearing on the through-flow capacity, while the kinetic energy imparted by a possible tangential velocity component is wasted.

A detailed hydromechanical and mathematical treatment of the theory of operation of axial turbines can be found in various reference books.[84–89] From the theory it follows that, under a certain head and at the rated speed, efficiency varies with the opening of the wicket gate which determines the discharge. Thus the efficiency curves can be plotted against discharge or, since the discharge is in definite relation to power, also against turbine output. By adequate turbine design the best efficiency can be allocated to the maximum output or, if necessary, to lower values as well.

If the runner blades are also adjustable (Kaplan turbine) the efficiency, in comparison with the propeller machines, remains very high over almost the entire output range. The efficiency curve can be plotted against discharge, gateage (percentage of the full-gate discharge), output, or percentage of the rated power. Figure 48 shows the efficiency curves of a Kaplan and a propeller wheel assuming that the designer allocated the maximum efficiency to 75% of the rated power. The significant difference in favour of the Kaplan machine can be easily understood when bearing in mind that the position of the runner blades (setting angle β) can be adjusted to any opening of the wicket gate, i.e. to any power output, to attain the highest possible efficiency. In other words, the efficiency diagram of a Kaplan machine is the enveloping curve of an infinite number of continuously adjoining efficiency curves of propeller wheels (see Fig. 49).

It follows from the above that fixed-blade (propeller) turbines are still justifiable when head and output (load) do not significantly vary. The above is also valid for all other types of axial machines (bulb, rim-generator, S-type), except that they do not require a spiral case.

In order to reduce expenses several modern high-powered stations are partly or completely equipped with normal propeller turbines, or even with turbines having the wicket gate fixed (see Fig. 41).

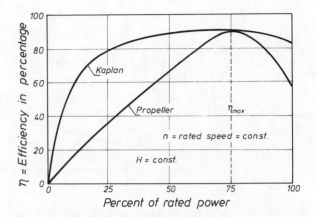

FIG. 48. Comparison of efficiency curves of axial turbines.

7.3.2 Cavitation and Static Draft (Suction) Head

The wheel can be exposed to cavitation when the absolute pressure drops below the vapour pressure prevailing at the actual temperature. From the Bernoulli theorem it follows that the lowest absolute pressure occurs in the exit section of the runner and, expressed in water column, is (Fig. 50):

$$\frac{p}{\gamma} = B - h_s - \frac{v_e^2 - v^2}{2g} + \Delta h \,[\text{m}] \qquad (36)$$

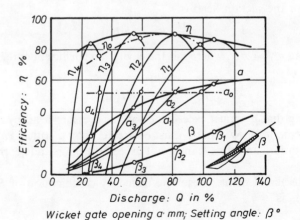

FIG. 49. Interpretation of the efficiency curve of Kaplan wheels.[89a]

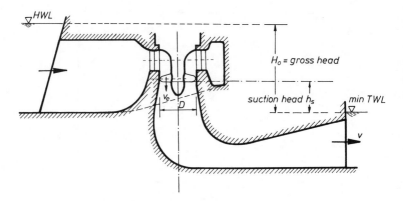

FIG. 50. Definition sketch for the static draft (suction) head.

where $B = p_0/\gamma$ denotes the actual barometric pressure and Δh the friction losses in the draft tube; h_s (the difference between the exit level of the runner and the TWL) is the section (static draft) head. If, due to an improper setting of the unit, $p/\gamma < p_v/\gamma$, where p_v denotes the vapour pressure, the lower tips of the runner blades, the runner hub and the lower portion of the discharge ring are exposed to cavitation. The pitting due to cavitation can, in extreme cases, completely destroy the blades and other parts of the wheel. Cavitation can also cause intolerable vibration, noise and considerable decrease in efficiency.

Neglecting kinetic energy and friction in eqn (36), the simplified expression for the suction head is

$$h_s = B - p/\gamma \, [\text{m}] \tag{37}$$

Substituting $p/\gamma = \sigma H$, where σ is the cavitation coefficient and H the net head, gives

$$h_s = B - \sigma H [\text{m}] \tag{38}$$

Considering that p/γ must exceed the vapour pressure head by a certain safety margin, it follows that σ must be greater than a critical value σ_c to avoid cavitation, and thus h_s must not exceed the magnitude corresponding to σ_c.

Since the actual plant σ resulting from any combination of H and h_s station data is

$$\sigma_p = \frac{B - h_s}{H} \, [\text{dimensionless}] \tag{39}$$

it is obvious that the design has to fulfil the condition $\sigma_p = \sigma_c$. By substituting the gross head H_0 into eqns (38) and (39), safe values for h_s and σ are obtained. It should be emphasized that for high-specific-speed turbines, especially when operating in the higher range of low-head, eqn (38) furnishes *negative* figures, which show how far the runner exit has to be submerged *below the minimum TWL* (see Figs 17, 18, 25). σ_c and σ_p are in close relation to the hydromechanical characteristics of the turbine and, in practice, determined as a function of its specific speed.

To find the decisive setting of the turbine, the permissible maximum h_s has to be related to the lowest TML, assuming the simultaneously possible highest HWL. Under some conditions, however, a sunk HWL is decisive for the setting height.

A more accurate approach for determining h_s can be accomplished by inclusion of the vapour pressure head[90] or also the dynamic draft head[91]

$$h_d = \frac{v_e^2 - v^2}{2g} - \Delta h \,[\text{m}] \tag{40}$$

in the derivation.

7.3.3 Operational, Specific, Synchronous and Runaway Speed

The speed of the revolving unit connected to a network is constant during its operation (disregarding the extremely slight fluctuation of the grid frequency, which has no bearing on the general design procedure and is of importance for dimensioning the governing system only). Selection of the operational (rated, normal) speed substantially influences the generator design, because the higher the speed the lower the number of rotor pole pairs required to satisfy the network frequency. Thus, according to eqn (27) and assuming $f = 50$ Hz, the necessary number of pairs of poles at a speed n (without speed increasing gear) is:

$$P_p = 3000/n \tag{41}$$

Further (see Section 8) with higher speed the same power output can be achieved with a lower-diameter rotor. Higher speed also entails a slightly lower turbine diameter. Summing up, the construction and installation costs of the entire powerhouse can be diminished by augmentation of the rated speed since, as is shown in Section 9, all main dimensions of the powerhouse vary in direct proportion to the machine diameters (turbine and generator rotor). Nevertheless, the scope for the selection of the rated speed is limited, as is shown below.

The geometrical configuration of the wheel (including also the prevailing position of the wicket gate and runner blades) defines the hydromechanical characteristics and operational behaviour of the turbine. Accordingly, wheels of different sizes but with geometrically similar shapes (homologous turbines), on the basis of the similarity laws, can be identified by a single parameter. In engineering practice specific speed is most commonly used:

$$n_s = n \frac{\sqrt{P}}{H^{5/4}} \text{ [rpm]} \tag{42}$$

where P is the rated power (full load) of the turbine in kW, H is the rated head in m, n is the rated speed in rpm. (It can also be related to peak efficiency.) From the derivation of this formula, the specific speed of a turbine can be formulated as the speed at which a geometrically similar (homologous) turbine would run under a head of 1 m if its size were selected so that it produced 1 kW power. The range of specific speeds for axial turbines is between 350 and 950 rpm. Earlier the specific speed was allocated to 1 HP, so the relation is

$$n_s = \sqrt{0 \cdot 736} \, n_{s(HP)} = 0 \cdot 857 n_{s(HP)} \tag{43}$$

If turbine data are in the foot-pound system the following conversion applies:

$$n_s = 3 \cdot 81 n_{s(ft\text{-}lb)} \tag{44}$$

At present, especially the turbine designers and manufacturers prefer dimensionless numbers (shape factors) to characterize the turbine types. Troskolanski was the first to introduce a dimensionless number,[92] though later various other type numbers were also suggested.[93]

In order to avoid cavitation within the runner (and also for other hydromechanical reasons) its velocity must not exceed a certain experimentally found part of the spouting velocity ($\sqrt{2gH}$). Hence, it can be proved that with increasing head the permissible specific speed has to be reduced. Several experts, manufacturers and institutions have presented formulas and diagrams displaying, for all types of turbines, the permissible highest n_s against rated head.[94,95] The author suggests[96] for conventional Kaplan turbines:

$$n_{s(max)} = \frac{2570}{\sqrt{H}} \text{ [rpm]} \tag{45}$$

which can be increased by about 7–12% for tubular machines (n_s related to kW).

Limit values of the cavitation coefficient σ for a safe design are given as a function of specific speed. Evaluation of *in situ* experiences and much research work has contributed to the clarification of this subject.[97-102] The author's recommendation for the first stage of planning for conventional Kaplan turbines is[103]

$$\sigma = \frac{n_s^{3/2}}{20\,000} \text{ [dimensionless]} \tag{46}$$

which can be lowered for propeller wheels by about 8–10%, and also reduced by a few % for tubular machines.

In most cases, it is economical to make full use of the limit given by eqn (46); under particular conditions, however, lower n_s and, accordingly, lower n values are selected. The author uses for the design based on the permissible maximum n_s the term 'limit design'.

Whatever the result of a calculation for operational speed, for the design the next synchronous speed has to be chosen as the unit can be connected to the grid only when it rotates with any of the speeds to be gained from eqn (27) by substituting whole numbers for p_p.

In an emergency caused by a sudden load rejection (i.e. when the generator is disconnected from the grid) the unit, assuming that the flow through the turbine is constantly and continuously maintained, accelerates up to the runaway speed. The highest runaway speed can develop at the highest head and full gate opening. At runaway speed the unit gets into a hydromechanical equilibrium and rotates without producing useful power. The runaway speed of axial machines may reach 2–3·5 times higher values than the rated speed. Since, in most cases, it is not reasonable to dimension the machines (both turbine and generator) and the bearings for runaway speed, various measures have been devised and successfully applied to prevent runaway (see Section 7.4.1).

7.3.4 Runner Diameter
Since the general design of the powerhouse, presuming that the number of units has been already decided, depends mainly on the setting elevation (defined by h_s) and the runner diameter, it is advisable to have an approximate estimate on the latter too.

Several proposals are known from the literature, expressing the runner diameter D alternatively as a function of Q, H, n, n_s (Ahlfors, 1926;

Pantell, 1933; Finnicome, 1940; Nesteruk, 1946; Berejnoy, 1948;[102] de Siervo and de Leva[95]). For the preliminary phase of planning the author suggests a simple relation valid only for the limit design:[103]

$$D = cH [\text{m}] \tag{47}$$

where the following values may be substituted for the diameter coefficient c:

(a) Conventional Kaplan turbines

$$c = 0 \cdot 40 - 0 \cdot 38 \ldots \text{if } H < 5 \text{ m}$$

$$c = 0 \cdot 39 - 0 \cdot 35 \ldots \text{if } H = 5 - 20 \text{ m}$$

$$c = 0 \cdot 37 - 0 \cdot 33 \ldots \text{if } H > 20 \text{ m}$$

(b) Tubular turbines

$$c = 0 \cdot 40 - 0 \cdot 35 \ldots \text{if } H < 5 \text{ m}$$

$$c = 0 \cdot 35 - 0 \cdot 31 \ldots \text{if } H > 5 \text{ m}$$

With more conservative solutions, i.e. when choosing significantly lower n_s and n values than those pertaining to the limit designs, the diameter is larger.

Very large Kaplan turbines have already been constructed, e.g. $D = 9 \cdot 5$ m (at Iron Gate, see Fig. 9).

Finally, it has to be repeated that the above recommendations concerning the limit magnitudes of h_s, σ, n_s and D are only general estimates for preliminary planning and feasibility studies. Manufacturers have their own designs, based on research and model tests, which have to be accepted for the final proposals.

7.3.5 Operation under Varying Head, Efficiency-Hill Diagram

If the head varies considerably, which is mostly the case with low-head plants, the determination of the operational characteristics of the turbine to be installed is not as simple as has previously been described, e.g. the efficiency curves in Figs 48 and 49 are valid for a single selected head only. With decreasing head efficiency diminishes and σ_c grows. The efficiency also drops when the head increases, but then σ_c decreases.

The performance characteristics of turbines over a wide range of discharge (or power output) and head can be obtained from the efficiency-hill diagrams, plotted on the basis of extensive tests on model turbines. The results are valid, with slight corrections due to scale effect, for homologous designs of any absolute dimension. Hence the performance of the prototype can be predicted from the operational characteristics or, in the case of dimensionless formulation, from the operational constants of the selected turbine geometry displayed by the hill diagram. Several versions of the hill diagram can be found. The characteristics and constants are derived by using the hydromechanical similarity laws. One procedure is presented below.

The so-called unit speed, unit discharge and unit power pertain to a particular (fictitious) unit-diameter (1 m) turbine homologous to the tested model and operating under unit head (1 m). These are denoted as n_{11}, Q_{11} and P_{11} and can be derived from the performance tests of the model. Between these unit characteristics and the corresponding quantities (n, Q, P) of any homologous prototype of diameter D operating under a head H, the following relationships exist:

$$n_{11} = \frac{n}{H^{1/2}} D \tag{48}$$

$$Q_{11} = \frac{Q}{D^2 H^{1/2}} \tag{49}$$

$$P_{11} = \frac{P}{D^2 H^{3/2}} \tag{50}$$

Usually n_{11}, Q_{11} or n_{11}, P_{11} are plotted. Figure 51 shows a plot in the (n_{11}, Q_{11}) system of the isolines of the following parameters of a Kaplan design: efficiency (dimensionless), wicket gate opening (a mm), runner blade setting ($\beta°$), cavitation coefficient (σ). Since, for homologous machines, efficiency and σ according to the strict similarity theorem are invariable, they can for the prototype be read directly from the diagram. Since, however, the accurate prediction of efficiency, especially of its maximum value, is of great importance, the correction (always an increment in favour of the larger prototype) is calculated by the manufacturer. Several formulae have been suggested for efficiency correction (taking into account the non-homologous friction losses).[104,105]

For a better understanding of the performance of the model tests, the evaluation of the hill diagram and the selection of the prototype operating point (n_{11} and n) further study of books dealing with this problem is recommended.[104,106,107]

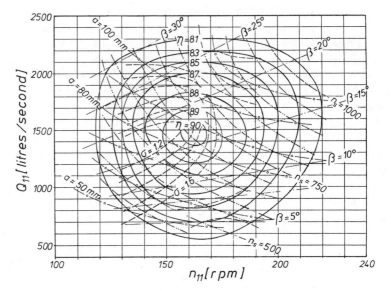

Fig. 51. Efficiency-hill diagram.

7.4 Governing

7.4.1 Flywheel Effect

The frequency of the electric power produced by the generator has to be maintained constant and, therefore, the balance between input torque and load torque must always be restored, if

(a) input torque changes because of variation in H or Q, or
(b) resistance moment of the generator changes because of variation in network load.

An ideal design of speed governor which would allow changing the input torque without delay, i.e. simultaneously with the load torque, cannot be realized for the following reasons:

1. A certain (even if very small) speed change is essentially the factor tripping the speed control process.
2. Inertia of moving masses (water and machines) introduces a delay in the follow-up of input torque.
3. Regulation time must necessarily be increased in order to limit water hammer shocks upon the structures and turbines.

The change in speed depends not only upon the velocity of governor response, but also upon the flywheel effect of the revolving parts of the turbine and generator. As the regulating device (i.e. the speed governor) cannot follow without delay fluctuations in load condition, a rather significant flywheel effect depending upon the sensitivity of regulation is required. The response speed of guide-vane control is limited, as mentioned under 3.

With small machines, where the flywheel effect of the rotating part of the aggregate is insufficient, special flywheels are employed. The usual procedure is that provision for the necessary flywheel effect is the task of the generator manufacturer, guided by the governor design of the turbine.

In the case of sudden load rejection, the time during which the runaway speed is reached is the longer the greater the flywheel effect. Since, in most cases, the machines are not designed for runaway speed, emergency installations are applied which, to avoid disturbance of operation, respond only to a speed clearly exceeding the highest value of the normal operational speed; some time elapses after governor response until the wicket gate can be completely closed, the unit still accelerates. Thus the necessary flywheel effect depends on the one hand on the turbine closure time and on the other on the permissible speed increment:

$$WD^2 \simeq 1800 \frac{PT}{\varepsilon n^2} \text{ [kNm}^2\text{]} \qquad (51)$$

where D is the diameter (m) of the generator runner, WD^2 is the flywheel effect (kN $\times$ m^2), P is the rated power (kW), T is the closure time(s) of the wicket gate, and $\varepsilon (= \Delta n/n)$ is the coefficient of the Δn permissible speed increment.

The relation of the flywheel effect to the inertia momentum I of the rotating mass is defined by

$$WD^2 = 4gI \qquad (52)$$

where g is the gravity acceleration. (It has to be borne in mind that W is not the weight of the rotating mass but that of a fictitious mass assumed to be distributed on an infinitesimally narrow strip along the perimeter of the rotor and satisfying eqn (52).)

There are also other time values besides the wicket gate closure time which may influence the determination of the flywheel effect (starting time of the water column, servomotor opening/closing time);[109] their discussion, however, is beyond the scope of this treatise.

The primary measure for preventing runaway is the emergency fast closure of the wicket gate during interval T (see above), actuated by the governor in the case of sudden load rejection. Sometimes planners insist upon a second safety measure, i.e. upon installation of a quick action emergency gate in the inlet flume, presuming the possibility of a simultaneous sudden load rejection and wicket gate failure.

7.4.2 Principles of Regulation and Components of the Governor

Since many devices in the electric power system require stable frequency (synchronous clocks, various types of electrical instruments) speed control of the generating units connected to the grid is essential and the generated power has to be adjusted to the power demand (load). The mechanism fulfilling this requirement is the governor. The speed sensing organ of the governor is either a rotating flyball pendulum or an electronic sensor.

The movement of the pendulum or the response of the electronic sensor is transferred by one-step or multi-step mechanisms to the servomotor actuating the wicket gate. Thus, by regulating the guide vanes the inflow to the runner is adjusted to the prevailing power demand.

In double-regulated (Kaplan) turbines, the position of the runner blades (setting angle β) is automatically co-ordinated with the wicket gate opening a (see Fig. 49) to provide for the maximum attainable efficiency. The required adjustment of the runner, defined by the cam-function, $\beta = f(a)$, can be achieved by mechanical or electronic transmission.

The governing system (actuator) includes the following devices: speed sensor, control valves, oil pressure tank, servomotor, regulating mechanisms for the wicket gate (cam adjustment for Kaplan machines). Hunting of the actuator must be avoided. (For further study of the governing system the reader is referred to the relevant literature.[109,109a,110–113]

7.4.3 Speed Adjustment of the Governor, Speed Droop and Load Sharing

When the load upon the network exceeds the input power supplied by all the power plants (hydro, thermal, nuclear) feeding the system, the frequency drops, while if the load drops below the input power the frequency begins to rise. Accordingly, in order to maintain frequency, the input power must be increased and diminished respectively. Some of the power stations have the task of participating in speed regulation.

The relation between speed and wicket gate opening, established by the mechanical or electronic sensor, is defined by an $n = f(a)$ function which can be transferred to an $n = f(P)$ characteristic, termed the statism of the governor. Linear statism is attainable by adequate design. The adjustment of the governor comprises two actions (Fig. 52(a)):

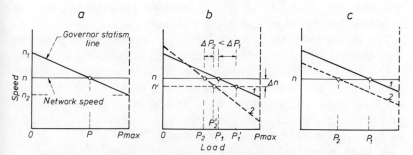

FIG. 52. Speed-droop diagrams.

1. Fixing the output (load) pertaining to the rated speed and the rated network frequency.
2. Selecting the speed range extending from the no-load condition (idle running) to full-load operation. This is the so-called speed-droop and is defined as

$$\delta = 2 \frac{n_1 - n_2}{n_1 + n_2} \text{ [dimensionless]} \tag{53}$$

Thus the droop is the ratio of the speed range to the mean value of the limiting speeds and expresses the gradient of the statism line (sensing speed).

From the above it follows that the flatter the statism line of a unit the greater its share in the change of load: $\Delta P = \Delta P_1 + \Delta P_2$ ($\Delta P_1 > \Delta P_2$); see Fig. 52(b). In other words, units with small speed-droop greatly respond to frequency changes. When the statisms of two units are parallel they produce differing powers at the network rated speed (or at any other speed), but their share is equal in any load change; see Fig. 52(c).

Consequently, in a power system, generating units (or complete plants) with very flat or almost zero speed-droop can be used as regulating units (plants) since they can pick up or drop load at the smallest change in frequency, whereas the other units (plants) with great speed-droop do not, or only slightly, react to load variations in the network. In base-load plants the δ can be as high as 0·04–0·06. Furthermore, other units (plants) supplying the power system can be operated by blocked load, i.e. by fixing their output in accordance with their optimum possible operating mode (e.g. firm power by nuclear plants) or complying with availability of their resources (e.g. natural discharge at hydropower plants or need to satisfy irrigation demands in multi-purpose schemes).

The adjustment of the governor has to be co-ordinated with efficiency characteristics of the turbines to operate them as efficiently as possible.

Low-head and particularly run-of-river plants are not generally used for frequency control. In order to utilize available river flow (up to the plant discharge capacity) the output has to be adapted to the natural hydrograph. This can be achieved by maintaining a constant HWL. Hence the governor is actuated by a water level sensor.[114,115]

A large power grid requires high-degree frequency stability. For example, in the German Federal Republic the current directives allow only a variation of 0·05 Hz of the standardized 50 Hz line frequency. Exceptionally, frequencies lower than 49·95 Hz occur and, according to the rate of frequency drop, various types of emergency measures are applied.[116]

8 ELECTRICAL EQUIPMENT

8.1 Generators and Transformers

The mechanical power output of the turbine is transformed into electric power by the generator (alternator). In hydro-power stations three-phase synchronous a.c. machines are generally employed. Single-phase a.c. generators are sometimes installed in plants serving special consumers (e.g. electric railway, electro-metallurgical plants). Sometimes d.c. generators are used to supply isolated industrial establishments.

The main parts of a generator are shown in Fig. 53. The thrust bearing of a typical vertical-axis hydro-generator (carrying the weight of the turbine runner, the axial component of hydrodynamic loads and the generator rotor) is mounted either on top of the head cover of the turbine, or on the lower or upper (see Fig. 53) bearing spider (supporting frame) of the generator.

The rotor holds on its circumference the poles excited by direct current. Salient and non-salient poles can be distinguished. The exciter current may be supplied to the coils of the poles by d.c. machines having a common shaft with the generator, or by exciter generators geared to the main shaft. The exciter can also be driven by an electric motor supplied from the network, or by a small turbine installed for this purpose. In some plants the d.c. exciting current is produced through rectifying a.c. current drawn from the network.

The terminal output (wattous power) of a three-phase generator, for

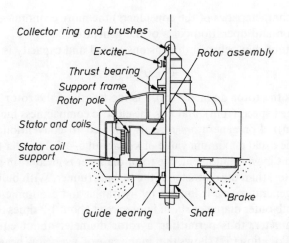

Fig. 53. Sketch of synchronous generator. (After C.C. Warnick.[117])

both star and delta connections, is given by

$$P = \frac{\sqrt{3}}{1000} UJ \cos \phi \, [\text{kW}] \tag{54}$$

where U is the mains voltage, J is the line current (amperes), ϕ is the phase shift angle, and $\cos \phi$ is the power factor. The usual rated voltage is between 3000 and 6000 V, while large units may be built for 10 000 V; generators with rated voltage of even 20 000 V or more exist.

The generator has to feed into the network the reactive (blind) power

$$P_r = \frac{\sqrt{3}}{1000} UJ \sin \phi \, [\text{kW}] \tag{55}$$

which is necessary to deliver magnetizing power to the network; this component, however, does not perform useful work. Consequently the generator proper has to be dimensioned from the apparent (nominal) power

$$P_n = \frac{\sqrt{3}}{1000} UJ = \frac{P}{\cos \phi} \, [\text{kVA}] \tag{56}$$

which is expressed in kilovolt-amperes. It can easily be proved that the higher the power factor depending on the electrical behaviour of the network (characteristics of the consumers) the more economical are both construction and operation costs of the generator.

The relationship between dimensions, speed and capacity is

$$P_n = \kappa D^2 \ln \text{ [kVA]} \tag{57}$$

where D is the rotor diameter (m), l is the length of the rotor body (m), n is the rated speed (rpm), and κ is the electric compactness factor (kVA/ $(m^3 \times$ rpm)). For generators directly coupled to conventional Kaplan turbines the usual maximum value of κ is about 5–5·5. In order to provide for sufficient flywheel moment, the rotor diameter is greater (in most cases much greater) than the diameter of the turbine runner. With bulb arrangement the generator size is limited by hydraulic and economic criteria so that the bulb outer diameter should not exceed about 1·2 times the turbine runner diameter, thus permitting a rotor diameter about equal to the runner.[118] Equation (57) shows that in the case of a very low head insertion of a speed increasing gear may be favoured for economic reasons.

Electric power systems have usually inductive (positive) resistance characterized by lagging power factor (the current lags behind the voltage), related to positive ϕ. Very exceptionally (e.g. during Sunday nights in winter) the opposite state in a grid may develop: capacitive resistance requiring condensers or generation performance with leading power factor, i.e. with negative ϕ.

The main sources of power losses in the a.c. generator are: d.c. power consumption for excitation and cooling, copper losses in the armature (stator) and rotor windings, friction and windage losses, core losses in the magnetic circuits. The overall efficiency of the generator mainly depends on its rated (nominal) power, on the relative load and on the power factor (Fig. 54).

In low capacity machines free-cooling systems are frequently used. In high-powered units, on the other hand, closed air coolers fed with fresh air or closed-circuit systems are installed.

The design of the generator brake is based on the operational torque and flywheel moment.

In small-capacity plants, to a great extent, horizontal axis (Fig. 19) or inclined axis outside (i.e. not bulb) generators are installed and, to reduce costs, they are frequently of the induction type.[117,119] The simpler induction generator has no excitation; it draws the requisite reactive power from the grid and, consequently, draws down the system voltage.

The electric current is conducted from the generator stator to the bus bar system by means of insulated cables or bus conductors, with the

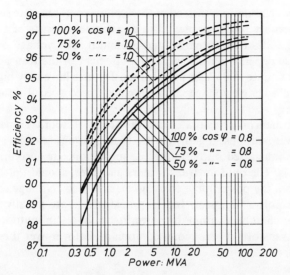

FIG. 54. Generator efficiency as a function of rated power, relative load and power factor.

insertion of power switches and disconnecting switches;[120,121] the circuit breakers are used to open or close electrical circuits under high-load currents, while the disconnecting switches are installed primarily to isolate dead lines or equipment from buses of live apparatus. Single and double bus systems can be distinguished. The double bus system used frequently in major plants ensures full freedom of switching operations during generator failures or other incidents. In large stations the buses and switches are placed in bus galleries.

Since the plants are usually situated at a great distance from the power-consuming centres, transformers to step up the generator voltage are required to reduce the $J^2 R$ heat losses in the transmission lines. If the bus system is run at generator voltage, any transformer pertaining to the plant may be connected to any generator. Another solution is for each transformer to form a coherent group with a generator (unit block system). Other combinations of connections exist.

Large transformers are filled with oil for insulation and cooling. Attention has been called to the fact that the great size and weight of high capacity transformers present transport problems which commonly surmount those of the plant itself.[121,122] In some high-power oil-immersed transformers the cooling oil circulation is forced by pumps or by fan activated air circulation. Instead of three-phase units separate single-phase transformers are often adopted. The transformer consists of a laminated sheet core and a coil system wound around it. The voltage on the high tension side depends on the special requirement of the transmission line (from 20 kV up to 400 kV or more). Transformers have very high efficiencies: 97–98%. High voltage circuit breakers need special design for arc extinction.

Transformers may be situated indoors or out-of-doors. Transformers of large power stations together with the high-tension-side switchgear, sometimes integrated into a multi-line substation, are accommodated in switchyards. The outdoor transformers, however, should not be situated too far from the powerhouse to avoid expensive low voltage connections and intolerable J^2R losses. Indoor transformers are usually located in galleries or chambers within the powerhouse. In a semi-outdoor arrangement the transformers are mounted upon a platform over the draft tube (Fig. 18).

8.2 Control, Protection and Emergency Outfit

Control of operation in the plant comprises the following main items: (a) machine starting and stopping; (b) automatic starting procedures; (c) synchronizing; (d) control of loading and frequency; (e) voltage control; (f) permanent supervision of machine running; (g) control of hydraulic parameters (HWL, flood control for opening the gates of the weir, discharge indication, etc.).[123] The indicating devices are usually concentrated in a separate control room. Optical and/or acoustical alarm systems are used. Adjacent plants, especially stations of a chain on the same river, can be operated by remote control.

Fire protection for both the generators and transformers is an essential feature of electrical design, since transformers represent a great fire hazard owing to the high quantity of stored oil. The protective system for hydrogenerators can be divided into two groups: (a) devices for preventing failures; (b) measures to eliminate defects caused either by insulation breakdown or by flashover across an insulation during operation. Transformers may have thermal protection by devices indicating temperature of winding and oil. Lightning protection is provided for the switchyard. The bus bar protection is mainly directed to earth-fault occurrence.

With the advent of very-high-tension switchgears and overhead transmission lines the various kinds of protection measures, and the construction of reliable indicators and alarm equipment, has become of paramount importance.

9 PRELIMINARY PLANNING OF THE POWERHOUSE

9.1 Powerhouse Equipped with Conventional Axial Turbines

Assuming that the power site, the general arrangement of the entire river barrage, the highest possible HWL have been fixed and the necessary hydrological data are available, the power curve for the station has to be established (see Fig. 36). The next step is an estimate on the number of generating aggregates. This can be made on the basis of experience or by optimization, i.e. by repeating the entire forthcoming procedure with diverse numbers of units and by comparing the investment costs.

When the unit capacity, the rated head and, from eqn (54), the highest permissible specific speed are known, the rated operational speed can be obtained from eqn (42). A recalculation with the next synchronous speed ensues. Further, eqn (47) presents the runner diameter. These are characteristics for the limit design. Based on experience and studies[95,124] the author suggests an informative sketch on the general arrangement of the complete system of passages with the limit values as shown in Fig. 55.[96]

The limit setting, i.e. the highest permissible level of the runner exit in relation to TWL, is established by using the Thoma equation (eqn (38)). If the head greatly varies and, particularly, if the lowest head is much smaller than the rated one, the setting level has to be checked for the limit heads too, since it may occur that the σ_{max} value at minimum head is so high that the $\sigma_{max} H_{min}$ product significantly exceeds the one pertaining to the rated head. Such a situation may require a lower setting, even when taking into account that this smaller h_s has to be related to a higher TWL (because, as shown in Fig. 36, the lowest heads develop at highest tailwater elevations). Similarly, the h_s evaluated from a cavitation coefficient σ_{min} and the highest head H_{max} has to be related to the lowest TWL.

The possibility of pondage (storage) has to be investigated as it may greatly augment the economic value of the plant.

The height of the draft tube exit section depends partly on the selected overflow velocity and partly on the criterion that the draft tube port must be submerged below the lowest TWL. The length of the intake flume is determined by the requirement on space for the installations within the passage. Following Fig. 11 from right to left the usual arrangement can be observed.

The intake section has to be sufficiently submerged below minimum HWL to avoid air intrusion. Sometimes an entrance sill is formed. In the case of large intake cross-sections the racks (screens) are strongly supported and care must be taken to achieve a vibration-free design.[125,126]

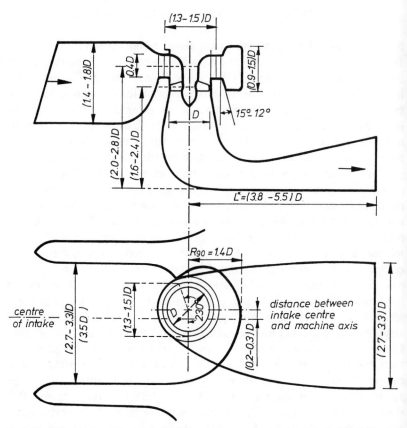

FIG. 55. Informative dimensions of the water passages of a conventional Kaplan setting (author's suggestion).

The hydraulic loss at the trashrack can be calculated by eqn (25). Reliable manual or mechanical cleaning of the rack is indispensable. Efficient raking machines are needed in up-to-date plants, especially if much floating debris can be expected.

Stop-log closure must be possible. In most cases a head-gate (intake gate) is installed between the stop-log slots and the spiral case. A single crane may serve both the stop-logs and the head-gate. For repair purposes stop-logs are also foreseen in the draft tube or at the end of it. In the case of very large dimensions the roofs of both passages (intake flume and draft tube) can be supported by dividing walls.[127–129]

The superstructure of the powerhouse protects the generator and other electrical equipment from rigours of weather. Indoor arrangement means the establishment of a machine hall with the major crane inside (Figs 11, 14–16, 24, 25). Generators of the semi-outdoor solution are housed in a deeper located low-roof gallery or chamber, while the crane moves outside, lifting machine parts through hatches in the roof (Fig. 21). In some plants no superstructure is erected at all and the generators are individually protected by weatherproof metal enclosures (Fig. 13).

Main auxiliary installations of the plants are: various types of bridge and gantry cranes, dewatering and drainage systems, water collecting sumps and pumps, lighting and ventilation systems, fire alarm and protection equipment, oil supply systems, heating and domestic water supply for the operating personnel.

Discussion of foundation methods and structural analysis of the various components of the powerhouse[130] are beyond the scope of this treatise. Avoidance of vibration and power swings is an important task of the designer.[131–134] Scale model studies for the proper layout of the entire river barrage and for the powerhouse bays are in most cases advisable.[53,135,136]

9.2 Powerhouse Equipped with Tubular Turbines

9.2.1 Bulb-Generator Setting

The first part of the procedure is similar to that discussed above, but there are slight differences as regards the estimate on permissible n_s, runner diameter D and static draft head h_s, as explained in Section 7 and obtainable from recent studies.[118,137–140]

Figure 56 illustrates a suggestion by the author for a rough estimate on main dimensions of the passages as a function of the runner diameter. From the tubular arrangement of the bulb unit it follows that the lowest TWL is always higher than the upper tip of the runner blade (negative suction head). [Data from the literature have been carefully checked since in many cases manufacturers and authors relate the TWL to the axis elevation of the runner and define the suction head (static draft head) accordingly.]

Rack and rack-cleaning devices are necessary but the headgate is usually dispensed with. Stop-log closures upstream and downstream of the machines are always provided for (e.g. see Fig. 17). Hatches and shafts have to be provided to facilitate dismantling the machines. Figure 57 shows three solutions: (a) in case of small bulb diameter a single shaft for both generator and turbine is sufficient; (b) a large-diameter bulb needs a

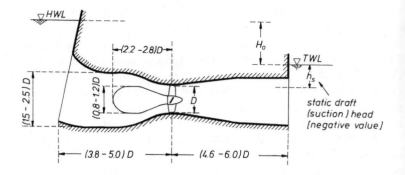

FIG. 56. Informative dimensions of the water passages of a tubular bulb setting (author's suggestion).

separate hatch for bulb and generator dismantling; (c) the pit arrangement provides the easiest way of handling the bulb. The interior of large bulbs is approachable by an access shaft.

9.2.2 Rim-Generator Setting

At present the author has no satisfactory data to present general guidelines for the rim-generator arrangement but, according to recent publications by experts from Escher Wyss, the length of this arrangement can be designed shorter than that of a powerhouse equipped with bulb machines, because the length of the inflow passage (i.e. the distance between rack and runner) can be reduced.[80,141,142] This advantage, however, can only be realized if the head is very low (see Fig. 46), while cases occur where the higher head and/or the requisite deep-setting of the runner govern the length of the inlet flume (e.g. in the powerhouse shown in Fig. 18).

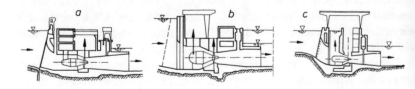

FIG. 57. Arrangements for the dismantling of bulb units.[118]

10 SMALL SIZE HYDROPOWER PLANTS

In Section 3.2.1 reference has already been made to 'small hydropower' (SHP). Within the SHP range limited by around 5 MW (or even 10 MW) some US publications distinguish two further sub-groups: plants are termed micro up to 100 kW and mini between 100 kW and 1 MW. The particular features and requirements respectively concerning small-size developments can be summarized as follows:

— Simplified hydrological and other physical studies which should be based, if possible, on regional analyses since the implementation of small plants can, from the economic point of view, be completely jeopardized by long-lasting, sophisticated and consequently too expensive investigations.
— Planning and construction needs highly qualified 'all round' engineers, capable of carrying out many professional tasks, possibly with only some short-term assistance from a few specialists and manufacturers.
— Use of local construction materials as far as possible.
— Use of standardized turbines. Individual turbine design is mostly too costly.
— The selection of simple (though not over-simplified) electrical equipment.
— The use of induction generators, if network conditions permit, may result in cost reduction too.
— In mini and micro plants 'package' machine units can also be installed advantageously.

We are witnessing extremely intensive worldwide progress in SHP development; thus SHP has become a very popular topic for national and international conferences. Since innumerable papers have recently been published on scientific hydrological, technological, ecological, economic and social aspects of SHP, the author instead of quoting individual treatises would like to refer to the proceedings of the most important meetings only.[143-146] Also a handbook[147] and specified chapters in textbooks on SHP furnish theoretical guidelines, design criteria and instructive examples for the planner.[148,149,149a]

11 ENVIRONMENTAL AND AESTHETICAL ASPECTS

It is necessary to supplement the plan, even at the first phase of its elabora-
tion, with an assessment on environmental feasibility. It is obviously not
an easy task to accomplish an objective analysis and a quantitative evaluation about the impact of the project on the environment. Although several
effects can, with sufficient physical, biological and other data, be more or
less acceptably forecast (e.g. influence on groundwater table, change in
water quality, impact on wildlife and fish), others can be evaluated only
qualitatively or very subjectively (e.g. impact on micro-clima, recreation,
landscape and architectural design).

Evidently a general statement that 'the structures of the project have to
harmonize with the landscape and must not disturb the scenic beauties of
the surroundings' would definitely comply with the justifiable requirements; nevertheless, what is 'harmonizing' and what is 'disturbing' are
fairly subjective judgements, though it can generally be stated that the less
that can be seen of the structures (i.e. the lower the superstructures and,
in the case of semi-outdoor or outdoor solutions, the more efficiently the
huge gantry cranes can be shifted into an inconspicuous position; see
Section 9.1) the better.

The partly or completely quantifiable physical and ecological impacts
have to be analysed in the preliminary phase of planning so that 'environmental surprises' are not encountered in the final stage of design, which
may enforce a complete re-planning of the project entailing an excessive
increase in costs, or even jeopardize the implementation.

Recently efforts have been made in several countries to improve environmental qualities of hydropower plants, and several measures have
been devised to diminish or even compensate for unfavourable environmental impacts.[150,151] As an example, the Lech River development in the
German Federal Republic may be quoted.[152] In order to replace biotopes
submerged in the headwater ponds, artificial islands and irregularly
shaped, low slope embankments without revetments were created.
Drainage and canals stabilizing the groundwater table were shaped to
simulate natural brooks. Submersible plants without any superstructure
are not conspicuous in dense forests. Low-roof powerhouses equipped
with tubular aggregates were designed to harmonize with the landscape.
The headwater reaches, where no conflict with nature preservation can be
presumed, are used for recreation and sport both in summer and winter.
Finally, an ancient diversion weir (probably constructed as early as about
1200 AD) has been reconstructed according to its original picturesque

design to re-establish the original fluvial landscape at the old city of Landsberg.

Because of the very widespread and sophisticated nature of environmental implications, checklists concerning all conceivable environmental aspects should be established.[153] In most cases evaluations of social, land-use, health and legal requirements are inseparable items of such an analysis.[154,155] The assessment process can be facilitated by using a methodology for systematizing, weighing and quantifying both detrimental and favourable environmental impacts.[156,157]

Small hydropower plants do not usually cause serious environmental problems; especially micro and mini developments may even offer ecological advantages for the watercourse and its surroundings. SHP plants can substantially contribute to erosion control in hilly and mountainous regions as worldwide destruction of forests could be stopped or at least decelerated by replacing firewood with hydraulic energy. According to the author's opinion a 100–120 kW midget plant can replace a metric ton of firewood per day.[158]

On the basis of the above considerations it seems to be justifiable to evolve specific methodology on environmental impact assessment for SHP developments.[159]

12 ECONOMIC APPRAISAL

12.1 Methods for Economic Evaluation

The economic analysis of a water power project is based on the comparison of all expenditures during its planning, design, construction and operation with the benefits obtained from the energy production over its so-called financial lifetime (amortization period). On the basis of experience over several decades, hydroelectric plants can usually be credited with an effective lifetime (availability until their obsolescence) substantially longer than their financial lifetime. Negligence of this fact not seldom leads to undervaluation of hydropower plants.

The current methods of economic evaluation are:

(1) Present-value (present-worth) assessment
(2) Internal rate-of-return evaluation
(3) Annuity method
(4) Benefit–cost ratio analysis
(5) Estimate on unit production cost

Since, during implementation of the project (preliminary studies, investigations and model tests, general planning, detailed designing, construction and installation), partial costs of the investment arise at diverse times and, similarly, the cash flow after commissioning manifests itself in annual OMR (operation, maintenance and repair) expenditures and annual benefits, it is imperative to convert all these cost and revenue items into equivalent monetary terms, i.e. to discount them to a selected common time by using a predicted interest (discount) rate and amortization period. The salvage value (buildings, lots, switching and sub-stations) should sometimes also be included as it may improve the economic parameters. On the other hand, it happens too often that the indispensable supplementary measures for environmental control are not, or not reliably, accounted for in the investment estimate, thus palliating the economic indicators.

Methods (1), (2), (3) and (5) may be termed as classical procedures of economic analysis, while the benefit–cost ratio (B/C ratio) was first suggested a few decades ago and at one time became so popular as to turn almost into a 'fashion' for evaluating engineering and other projects. In the author's opinion the B/C ratio is not a true economic characteristic and, besides, it can be misleading because its mathematical structure makes easier a broader manipulation than is rendered possible by application of the other methods.[160] Still, the B/C ratio should not be excluded for it can efficiently be used for preliminary screening of project variants, provided that common evaluation procedures are used. The reader may find the detailed description of the calculation processes of methods (1) and (4) in any textbook on economics[161–163] and also in hydraulic engineering publications.[160,164–166]

In some types of industrial enterprises, and also with hydropower plants, the fifth economic indicator, i.e. the unit production cost, is preferred for decision making, i.e. the overall production cost of *one* kilowatthour of energy is determined. The annual cost consists of the fixed (first) cost and the running expenses (OMR). Assuming a uniform annual repayment over the n-year amortization period, the so-called capital recovery factor

$$a^* = \frac{(1 + i)^n i}{(1 + i)^n - 1} \; [1/a] \tag{58}$$

including both capital repayment and interest charge, is calculated where i is the interest rate. Sometimes (mainly in Europe) the $q = 1 + i$ substitution is used, so that with the q interest factor eqn (58) appears as

$$a^* = \frac{q^n(q - 1)}{q^n - 1} \ [1/a] \tag{59}$$

After uprating all investment costs to the beginning of the year in which the presumed n-year power production commences, and thus obtaining an I \$ integrated present-worth investment, the annual fixed cost, according to the sinking-fund depreciation procedure, equals

$$C_1 = a^*I \, [\$/a] \tag{60}$$

Denoting the annual running cost (OMR), including management and taxes, by C_2 and, at least in the first approach, supposing its uniform distribution, the total annual cost to be charged to the yearly power production equals $C_1 + C_2$ dollars. Accordingly, the unit production is

$$C_0 = 100 \, \frac{C_1 + C_2}{E} \ [\text{cent/kWh}] \tag{61}$$

where E indicates the producible kilowatt-hours in an average year.

The experienced planner, in possession of some hydrological data, is often able to make a reliable estimate on the utilization hours t_u of run-of-river schemes (see Section 5.3 and Fig. 39). Thus, considering a rated plant capacity of P kW, the annual energy, according to eqn (34), will roughly total to Pt_u kWh, and eqn (61) can be expressed as

$$C_0 = 100 \, \frac{C_1 + C_2}{Pt_u} \ [\text{cent/kWh}] \tag{62}$$

It has to be emphasized that the calculation of the capital recovery factor according to the above extremely simplified procedure, i.e. by assuming a constant n value for the complete investment, is only acceptable if the planner has sufficient experience for assessing a fairly reliable fictitious amortization period being representative for the entire project. Otherwise, as is common practice, diverse amortization periods have to be assigned to the various elements of the plant. The heavy civil engineering parts (concrete and reinforced concrete structures, canals with their revetments, stone- and earth-works) have, assuming proper quality, a much longer life than the steel structures, machines and electrical equipment. Since, however, it is not possible to appraise the different n periods by an exact calculation procedure, the planner has to rely upon guidelines or directives issued by competent authorities or major hydroelectric enterprises. A concise compilation of some informative data collected from relevant publications is given in Table 5.[167,168] Since the result of the economic analysis is very sensitive to the choice of both interest rate and amortization period, it is obvious that the fairly wide ranges in Table 5 involve a high degree of uncertainty and can be accepted as rough information only. In every individual case the specified guidelines depending on the features of the relevant capital market and monetary policy have to be adopted. This principle is also valid when choosing the discount rate to be used for the planning.

It is usual to estimate, on the basis of long experience, the annual OMR cost as a function of the entire investment cost (capital):

$$C_2 = bI [\$/a] \tag{63}$$

where the running cost factor $b[1/a]$ roughly varies between 0·005 and 0·025 (0·5–2·5%), the lower values pertaining to high plant capacities and vice versa.

The capital demand, and in a very approximate way, the economic attractiveness of power plants is usually characterized by the specific investment:

TABLE 5

RANGES OF AMORTIZATION PERIODS FOR ECONOMIC ANALYSIS OF HYDROELECTRIC PLANTS[a]

Nature of asset	Period (years)
Heavy earth, rock, stone, brick, concrete and reinforced concrete structures (canals, tunnels, caverns, dams, weirs, intakes, etc.)	50–100
Timber structures with proper impregnation (sluices, gates, flumes, etc.)	20–25
Superstructures (machine halls, service buildings, dwelling houses, etc.)	35–80
Steel structures (gates, cranes, racks, etc.)	20–40
Machinery (turbines, generators, transformers, switchgear, auxiliary equipment, cables, etc.)	20–45
Transmission lines	20–25
Preliminary investigations	50

[a] Based on guidelines adopted by various government authorities and national power boards.

$$I_0 = I/P \, [\$/kW] \tag{64}$$

i.e. the capital required for *one* kilowatt installed capacity.

Substituting eqns (60), (63) and (64) into eqn (62) we obtain

$$C_0 = 100(a^* + b) \frac{I_0}{t_u} \, [\text{cent/kWh}] \tag{65}$$

The degree of utilization of a power site (head and/or plant design flow) and the number of generating units are sometimes established by an optimization process. Nevertheless, a reliable optimization needs fairly accurate physical, technological data and realistic assumptions of various types of costs and, especially of the discount rate and amortization period, since otherwise the result (optimum value) may be pointless and misleading. To avert such a mistake a sensitivity analysis is essential. (In the words of L.D. James and R.R. Lee: 'Benefit-cost analysis can be and has been corrupted'.[169])

Finally, a brief reference to multi-purpose hydroelectric schemes has to be made. A detailed discussion of this topic, however, is beyond the scope of this presentation. From a financial and economic point of view it is in most cases a very sophisticated task for the planner to perform the allocation of the investment and OMR costs to the diverse project objectives. Various cost allocation procedures, resulting in very different cost distributions, have been suggested.[160,164,170,171] These methods have a common element: the so-called separable costs must first be set aside, i.e. those costs have to be evaluated which are clearly chargeable to every single project purpose. Thereafter, the so-called joint cost has to be distributed according to a—mostly subjectively—selected allocation principle. From among the numerous proposals two are named here:

(a) The remaining-benefit method
(b) The alternatively justified expenditure analysis

In many cases, however, economical concepts are not decisive for the decision maker since political and/or social aspects also govern the allocation of the investment.

12.2 Informative Data

It is not an easy task to collect data on investments and actual energy production costs. Therefore, any comprehensive publication concerning actual financial and economic features of projects deserves credit.[7,172]

Because of the obviously enormous scattering of the investments, and due to the significant escalation of prices in time, it does not seem appropriate to present cost data of general validity. Besides, as Gordon pointed out,[172] the costs at newer sites will be higher than those at previous sites, because the most economic sites will be exploited first. Based on his former evaluations, Gordon derived empirical formulae for average new sites. In general, it can be stated that the specific investment cost (I_0) diminishes with growing head and with increasing capacity. According to a diagram by Gordon, the specific investment cost in the low-head region, provided that optimum site conditions exist, can roughly be estimated as follows (at 1982 prices):

$$\text{for } P \simeq 100 \text{ MW at } 10 \text{ m head} \ldots \sim 2100 \, \$/\text{kW}$$
$$50 \text{ m head} \ldots \sim 1400 \, \$/\text{kW}$$
$$\text{for } P \simeq 50 \text{ MW at } 10 \text{ m head} \ldots \sim 2400 \, \$/\text{kW}$$
$$50 \text{ m head} \ldots \sim 1600 \, \$/\text{kW}$$
$$\text{for } P \simeq 10 \text{ MW at } 10 \text{ m head} \ldots \sim 3100 \, \$/\text{kW}$$
$$50 \text{ m head} \ldots \sim 2100 \, \$/\text{kW}$$

Let us evaluate the unit production cost for a high capacity run-of-river plant constructed on a navigable river (e.g. on the Danube), providing good site conditions. The total investment in the river barrage, also including expenditures for all necessary structures along the backwater reach, can be estimated on a current price level as 550×10^6 US$. The main design data of the project are: $Q_p = 3150 \, \text{m}^3/\text{s}$, $H = 11 \, \text{m}$, $P = 300 \, \text{MW}$ (nine bulb-Kaplan units at $34 \cdot 4 \, \text{MW}$), $E = 1720 \times 10^6$ kWh/a.

The specific investment in the entire double-purpose project amounts to $(550 \times 10^6)/(300 \times 10^3) = 1830 \, \$/\text{kW}$. In cases when, on rivers with heavy traffic, it is assumed that the continuity and safety of navigation (with special regard to the low-water periods) can be enhanced by construction of the barrage, it is usual for the government to meet a significant portion of the expenditure by allocating it to navigation, i.e. investment and OMR costs of the ship lock(s) and possibly a certain part of the weir and of the measures in the backwater reach. In the case under discussion it would be 30%. Thus, the specific investment cost to be charged to power generation is reduced to $1280 \, \$/\text{kW}$. The capital recovery factor has been assumed at 8%, the running cost factor 2%. Hence $a^* + b = 0 \cdot 10$. By substituting this value and the annual utilization hours

$$t_u = \frac{1720 \times 10^6 \, [\text{kWh/a}]}{300 \times 10^3 \, [\text{kW}]} = 5740 \, [\text{h/a}]$$

FIG. 58. Downstream view of the Birsfelden low-head plant on the High Rhine (layout on Fig. 5). The photograph shows the high-superstructure powerhouse equipped with conventional Kaplan units and a plant of the gated weir. (Courtesy of E. Fabian.)

into eqn (65) the production unit cost will be

$$C_0 = 100 \times 0.1 \frac{1280}{5740} = 2.24 \, [\text{cent/kWh}]$$

Equation (65) shows that the unit production cost is very sensitive to the load factor and annual utilization hours. Low-head run-of-river developments have often extremely low load factors. To display this effect let us presume for a similar plant as analysed above a load factor as low as 0.4. Thus, according to eqn (34), utilization hours are $0.4 \times 8760 \simeq 3500$ [h/a] and the unit production costs augment to

$$C_0 = \frac{5740}{3500} 2.24 = 3.68 \, [\text{cent/kWh}]$$

Gordon also gives informative values on specific investment costs for small-size (SHP) plants (between 1 and 10 MW). In SHP, however, one has to differentiate between new projects and plants which are associated with refurbishment of old ones. In Europe much effort has recently been made to renovate and resuscitate SHP stations which became obsolete or had been put out of service during the 'oil boom'. Also, numerous SHP plants as additional structures are under construction at existing small dams created earlier for other purposes (e.g. for irrigation). Outstanding results have been achieved in this field in China (76 000 SHP stations, with a total installed capacity of 8500 MW constructed up to the end of 1983).[173] Obviously, it makes a great difference in expenditure which type of SHP

FIG. 59. Downstream view of the Säckingen low-head plant on the High Rhine; the semi-outdoor powerhouse is equipped with conventional Kaplan units. (From a prospectus on the High Rhine Development, by courtesy of the Kraftwerk Laufenburg.)

FIG. 60. Aerial view of the entire low-head river barrage Geisling on the Upper Danube: powerhouse, weir and lock. Conventional Kaplan sets are installed in the powerhouse. (Reproduced, with permission, from a prospectus of the Rhein-Main-Donau AG, München.)

FIG. 61. The Urspring low-head plant on the Lech; the low-roof semi-outdoor powerhouse accommodates bulb-generator Kaplan aggregates. (Reproduced, with permission, from a prospectus of the Bayerische Wasserkraftwerke AG, BAWAG, München.)

plant we are referring to: completely new projects, additional plants, rehabilitation or resuscitation of old stations. The scattering of the specific investment costs for mini and micro plants is extremely wide. Plenty of useful data can be gained from the innumerable case studies presented in the conference proceedings quoted previously and in several issues of the periodical *International Water Power and Dam Construction*.

In conclusion of this chapter, a few photographs of typical-up-to-date low-head developments are presented (Figs 58–61).

REFERENCES

1. MOSONYI, E. *Water Power Development*, Vol. I, 2nd English edn, Publishing House of the Hungarian Academy of Sciences, Budapest, 1963, Chapter 1.
2. QASIM, S. H. *SI Units in Engineering and Technology*, Pergamon Press, Oxford, 1977.
3. SOCRATES, G. and SAPPER, L. J. *SI and Metrication Conversion Tables*, Newnes-Butterworths, London, 1969.
4. BROWN, F. (ed.). *Statistical Year Book No. 7*, World Power Conference, London, 1954.

5. Reference 1, Chapter 5.
6. COTILLON, J. L'hydroélectricité dans le monde (Hydroelectricity in the world), La Houille Blanche (France), special issue 1978/nos 1–2.
7. GORDON, J. L. Recent developments in hydropower, *International Water Power and Dam Construction* (UK), September 1983.
8. MERMEL, T. W. Major dams of the world—1985, *Int. Water Power Dam Constr.*, July 1985.
9. SLEBINGER, V. Statistics of all existing waterpower resources, *Trans. 4th World Power Conf.*, London, 1952.
10. BLOSS, W.H. *et al.* Survey of energy resources 1980, 11th World Energy Conf., Munich, 1980, University of Hannover, Fed. Inst. for Geosciences and Natural Resources.
11. MOSONYI, E. *Water Power Development*, Vol. I, 3rd English edn, Publishing House of the Hungarian Academy of Sciences, Budapest, 1987, Chapter 6.
12. KARADI, G. *et al.* (eds). *Proc. Int. Conf. Pumped Storage Development and its Environmental Effects* (University of Wisconsin-Milwaukee, September 1971), Amer. Water Resources Assoc., Urbana, Ill., USA, 1971.
13. GRAY, T. J. and GASHUS, O. K. (eds). *Tidal Power* (Proc. Int. Conf. Atlantic Industrial Research Institute, Nova Scotia Techn. College, Halifax), Plenum Press, New York/London, 1972.
14. WILSON, E. M. and BALLS, M. Tidal power generation. In: *Developments in Hydraulic Engineering*, Vol. 4, ed. P. Novak, Elsevier Applied Science Publishers, London, 1987.
15. Reference 1, Chapter 4.
16. CASACCI, S. Large bulb units for tidal powerplants, *Int. Water Power Dam Constr.*, June/July 1978.
17. FOCAS, D. Straflo turbines for tidal application, *2nd Int. Symp. Wave and Tidal Energy*, British Hydromechanics Research Association, Cambridge, England, 1981.
18. CLARK, R. H. Re-assessing the feasibility of funding tidal power, *Int. Water Power Dam Constr.*, June 1978.
19. BERGE, H. (ed.). Wave energy utilization, *Proc. 2nd Int. Symp. Wave Energy Utilization*, Norwegian Inst. Technology, Trondheim, 1982.
20. Reference 11, Chapter 4 (subdivision: Wave Energy).
21. Nutzung der Wellenenergie (Utilization of wave energy), *Proc. Congr. 'Wasser Berlin 1981'*, Institut für Wasserbau und Wasserwirtschaft.
22. DAVIES, P. G. (ed.). *Wave Energy* (Department of Energy R & D programme 1974–1983), HM Stationery Office, London, 1985.
23. Reference 11, Chapter 4 (subdivision: Depression Power Plants).
24. BASSLER F. New proposals to develop Quattara Depression, *Int. Water Power Dam Constr.*, January 1977.
25. Israel will Mittelmeerwasser in das Tote Meer leiten (Israel intends to convey water from the Mediterranean into the Dead Sea), *Verein Deutscher Ingenieure Nachrichten (VDI News)*, No. 37, September 1980.
26. Reference 1, Chapter 34.
26a. BÖHMER, H. Das Donaukraftwerk Aschach (Danube power plant Aschach), *Wasserwirtschaft* (Fed. Rep. Germany), 1962/8–9.

26b. KÖNIGSHOFER, E. Aschach station, *Int. Water Power Dam Constr.*, August 1962.

26c. AEGERTER, A. and BOSSHARDT, A. G. (Consulting Engineers). Das Kraftwerkprojekt Birsfelden (Power project Birsfelden), *Schweizerische Bauzeitung*, 1949/37.

27. ROEHLE, W. Das Donaukraftwerk am Eisernen Tor (Danube powerplant at the Iron Gate), *Wasser Energie Luft* (Wasser- und Energie Wirtschaft, Switzerland), special issue, No. 3/4.

28. PARTL, R. VON. Das Donaukraftwerk Ybbs-Persenbeug (Danube power plant Ybbs-Persenbeug), *Zentralblatt für die Österreichische Industrie und Technik*, 1946.

28a. BÖHMER, H. Donaukraftwerk Ybbs-Persenbeug: Entstehungsgeschichte und Zusammenfassung (Danube plant Ybbs-Persenbeug: history and summary), *Österreichische Zeitschrift für Elektrizitätswirtschaft* (Austria), 1955, no. 9.

28b. GRZYWIENSKI, A. *Flusskraftwerke und Stromwerke (Single-Purpose River Power Plants and River Barrages on Navigable Streams)*, Springer, Vienna, 1948.

29. AIMONT, E. La centrale hydroélectrique de Monsin-lez-Liège, *Technique des Travaux*, 1954, nos 3–6.

30. SEIFERT, H. Die Mainstufe Grossmannsdorf (Main river barrage G.), *Bautechnik*, 1956, no. 6.

31. KEYL, L. and HAECKERT, H. *Wasserkraftmaschinen und Wasserkraftanlagen (Turbines and Hydroplants)*, Koehler Press, Stuttgart, 1949.

32. RAMSAHOYE, S. I. Calculating load rejection surges, *Int. Water Power Dam Constr.*, March 1978.

33. CÁBELKA, J. and GABRIEL, P. Inland waterways. In: *Developments in Hydraulic Engineering*, Vol. 3, ed. P. Novak, Elsevier Applied Science Publishers, London, 1985.

34. *La Première Chaine des Centrales du Rhin (First Chain of Stations on the Rhine)*, Electricité de France, 1959.

35. RUSSO, G.A. Volga development, *Int. Water Power Dam Constr.*, May/June 1958.

36. *Ausbau des Rheins zwischen Kehl/Strassburg und Neuburgweier/Lauterburg (Rhine Development between K/S and N/L)*, Prospectus, Rheinkraftwerk Iffezheim GmbH, 1976.

37. SCHAUFELBERGER, W. Das Mur-Kraftwerk Weinzödl (Weinzödl power plant on the Mur river), *Wasserwirtschaft*, 1982, no. 5.

37a. VAN VRANKEN W. P. Tube turbines to modernize hydro plants, *Allis-Chalmers Engng Review*, 1969.

38. ROUVE, G., Untersuchungen über den Krafthaus-Trennpfeiler (Investigations on the dividing wall of the powerhouse), *Wasserwirtschaft*, 1960/4–5.

38a. Reference 11, Chapter 34.

38b. WITTMANN, H. and GARBRECHT, G. Untersuchungen für Laufwasserkraftstufen (Investigations on run-of-river hydropower plants), *Wasserwirtschaft*, November 1954.

39. GRENGG, H. Ausbau der Drauwasserkraft und das Pfeilerkraftwerk (Development of hydropower of the Drau river and the pier-head layout), *Zeitschrift des österreichischen Ingenieur- und Architekten Vereins*, 1947/23–24.

40. GRENGG, H. and LAUFFER, H. Das Kraftwerk im Strom, Pfeilerkraftwerk (Power station in the stream, pier-head plant), Österreichische Wasserwirtschaft, 1949/9–10.

41. OBERLEITNER, P. Das Strömungsbild des Pfeilerkraftwerkes (Flow pattern at the pier-head station).

42. SCHULZ, A. B. Eine neue Pfeilerkraftwerksbauart (New type of pier-head plant), Wasserwirtschaft, 1949–50/2.

43. STRAUCH, A. B. Entstehung, Verhütung und Beseitigung von Eis in stehenden und fliessenden Gewässern und insbesondere an Stauanlagen (Formation of, protection against and disposal of ice in still and flowing waters and especially at river barrages), Deutsches gewässerkundliches Jahrbuch (Koblenz), special issue, 1954, No. 10.

44. Reference 1, Chapter 36.

45. FENTZLOFF, H. E. Principes fondemantaux de la construction des centrales submersibles (Basic principles for construction of submersible stations), Houille Blanche, 1949/5.

46. LEUTELT, H., Über die Anwendung von Ejektor-Lehrschüssen im Turbinenblock von Niederdruck-Wasserkraftwerken (Ejector outlets in low-head powerhouses), Wasserwirtschaft, 1960/11.

47. McQUEARY, F. K. and LEPPIN, W. Hydro power in Western Germany, Int. Water Power Dam Constr., May/June 1949.

48. Reference 11, Chapter 37.

49. UEDA, T. Large capacity bulb units in Japan, Int. Water Power Dam Const., March 1983.

50. HABERMAAS, F., Geschiebeeinwanderung in Werkkanäle und deren Verhinderung (Bedload intrusion into power canals and preventive measures), Wasserwirtschaft, 1935.

51. Reference 1, Chapter 21.

52. TISON, L. J. Au sujet de certaines dispositions permettant de réduire les quantités de matières solides entraînées par un système à surface libre (Measures for reducing sediment entrainment in free-surface systems), Compte Rendu des Deuxièmes Journées de l'Hydraulique, Société Hydrotechnique de France, Houille Blanche, 1952.

53. NOVAK, P. and ČÁBELKA, J. Models in Hydraulic Engineering, Pitman Advanced Publishing Program, 1981.

54. SCHEUERLEIN, H. Die Wasserentnahme aus geschiebeführenden Flüssen (Diversion in Sediment Carrying Streams), Ernst & Sohn, Berlin, 1984.

55. Reference 1, Chapter 23.

56. KIRSCHMER, O. Untersuchungen über den Gefälleverlust an Rechen (Investigation on head loss at racks), Mitteilungen des Hydr. Inst., T. H. München, 1926/1.

57. FELLENIUS, W. Undersökningar beträffende fallförluster i skyddsgrindar vid vattenkraftanläggningar (Investigation of Rack-Losses at Water Power Stations), Publ. University of Stockholm, 1927.

58. ESCANDE, L. Hydraulique Générale, Paris, 1943.

59. ZIMMERMANN, J., Widerstand schräg angeströmter Rechengitter (Hydraulic Resistance of Racks Caused by Oblique Inflow), Institute for Hydraulic Structures and Agricultural Engineering, University of Karlsruhe, 1969, No. 157.

60. GRAF, W. H. *Hydraulics of Sediment Transport*, Part 3, McGraw-Hill, New York, 1971.
61. ZANKE, U. *Grundlagen der Sedimentbewegung (Principles of Sediment Movement)*, Springer, Berlin, 1982, Chapter 9.
62. Reference 11, Chapter 27.
63. DELATTRE, P. and HENRY, M. The harnessing of the Rhine (extrait de Hydraulique et Electricité Française), *Houille Blanche*, 1951.
64. GENTILE, G. Note sul rivestimento in conglomerato cementizio dei canali in terra per gli impianti idroelettrici di Vizzola, Tornavento, Turbigo, Cimema e Pontey (Concrete revetments in the earth canals of the hydroelectric plants V, T, T, C and P), *Energia Elletrica* (Italy), 1951/5.
65. RAJARATNAM, N. Surges of finite height, *Int. Water Power Dam Constr.*, November 1982.
66. WITTMANN, H. and BLEINES, W. Kraftwerkschwalle und Schiffahrt (Power station surges and navigation), *Schweizerische Bauzeitung*, 1953/34.
67. ANNEMÜLLER, H. Der Einfluss des Turbinenschnellschusses auf die Schiffahrt in Kraftwerkstrassen (Effect of sudden turbine closure on navigation), *Bauingenieur*, 1960/9.
68. Reference 53, Chapter 8, Section 8.3.
69. HAGER, M. Der Oberrheinausbau und das Kulturwehr Kehl/Strassburg (Upper Rhine development and the control weir K/S), *Jahrbuch der Hafenbautechnischen Gesellschaft*, Vol. 39, Springer edition, 1982.
70. FELKEL, K. Die Geschiebezugabe als flussbauliche Lösung des Erosionsproblems des Oberrheins (Erosion control of the upper Rhine by artificial sediment supply), *Mitteilungsblatt der Bundesanstalt für Wasserbau* (Karlsruhe, FRG), No. 47, June 1980.
71. SHELDON, L. H. and RUSSEL, G. J. Determining the net head available to a turbine, *Trans. 2nd Symp. Small Hydropower Fluid Machinery*, Phoenix, Arizona, November 1982.
72. WARNICK, C. C. *et al. Hydropower Engineering*, Prentice-Hall, Englewood Cliffs, New Jersey, 1984, Chapter 2.
73. DIAMANT, Y. Y. and HERAPATH, R. G. Computer optimization for run-of-river energy, *Int. Water Power Dam Constr.*, November 1983.
74. MOSONYI, E. Hydro-electric development in Hungary, World Power Conf., Rio de Janeiro, 1954.
75. MOSONYI, E. and SCHILLING, F. Effect of river canalization on navigation, 19th Int. Navigation Congr., London, 1957.
76. FUCHS, H., Die Donaustufe Jochenstein und ihr Einfluss auf die Schiffahrt (Danube plant Jochenstein and its influence on navigation), *Binnenschiffahrtsnachrichten*, 1956.
77. MOSONYI, E. Studies for intake structures of flood detention basins to be established along the Upper Rhine, *Proc. Int. Conf. Hydraulic Aspects of Floods and Flood Control*, Brit. Hydromech. Res. Assoc., London, September 1983.
78. FELKEL, K. Vergleichende Gegenüberstellung von Kippbetrieb und gestaffeltem Hochfahren der Durchlaufspeicherung von Flusskraftwerken (Comparison between tilting and heaving peak operation of stream plants), *Wasserwirtschaft*, 1957/12.

79. Reference 11, Chapter 13.
80. HÖLLER, K. and MILLER, H. Bulb and Straflo (R) turbines for low head power stations, *Escher Wyss News* (Switzerland), 1977/2.
81. MOSONYI, E. *Water Power Development*, Vol. II, 2nd English edn, Publishing House of the Hungarian Academy of Sciences, Budapest, 1965, Chapter 114.
82. OSSBERGER-TURBINENFABRIK (Prospectus), Weissenburg, FRG.
83. HAIMERL, L. A. Die Durchströmturbine (The cross-flow turbine), *Energie*, Volume 11, Number 8.
84. LUNDGREN, G. H. and STILLE, C. E. Kaplan and propeller turbines. In: *Hydro-Electric Engineering Practice* (handbook in three volumes, ed. J. Gutherie Brown), Blackie, London/Glasgow, 1958, Vol. 2, Chapter 4.
85. Reference 1, Chapters 45–53.
86. BROWN, L. P. and WHIPPEN, W. E. *Hydraulic Turbines*, Parts 1 and 2, Pa. Int. Textbook Company, 1972/1977.
87. DAILY, J. W. *Hydraulic Machinery: Engineering Hydraulics*, ed. H. Rouse, John Wiley and Sons, New York, 1950, Chapter 13.
88. CSANADY, G. T. *Theory of Turbomachines*, McGraw-Hill, New York, 1964.
89. RAABE, J. *Hydro Power*, Verein Deutscher Ingenieure Verlag, Düsseldorf, 1985, Chapters 5–7.
89a. KEYL, L. and HÄCKERT, H. *Wasserkraftmaschinen und Wasserkraftanlagen (Turbines and Water Power Plants)*, Koehler, Stuttgart, 1949.
90. Reference 72, Chapter 7.
91. Reference 1, Chapter 51.
92. Reference 1, Chapter 47.
93. Reference 89, Section 9.2.3.
94. Reference 72, Chapter 4, Fig. 4.3.
95. SIERVO, DE F. and LEVA, DE F. Modern trends in selecting and designing Kaplan turbines, *Int. Water Power Dam Constr.*, December 1977.
96. Reference 11, Chapter 57.
97. US Department of the Interior. *Selecting Hydraulic Reaction Turbines*, Engineering Monograph No. 20, Bureau of Reclamation, Denver, 1976.
98. J. M. VOITH GmbH. *Cavitation-Setting and Definitions*, Heidenheim, October 1976.
99. LINDESTROM, L. E. *Review of Modern Hydraulic Turbines and Their Application in Different Power Projects*, Karlstads Mekaniska Werkstad (KMW), Kristineham (Sweden).
100. KNAPP, R. T., DAILY, J. W. and HAMMITT, F. G. *Cavitation*, McGraw-Hill, New York, 1970.
101. Reference 72, Chapter 7, Figs 7.5 and 7.8.
102. BEREJNOY, A. A., *Design of Water Power Stations and Operating Equipment* (in Russian), Leningrad/Moscow, 1948.
103. Reference 11, Chapter 5.
104. Reference 72, Chapter 4.
105. HUTTON, S. P. Component losses in Kaplan turbines and prediction of efficiency from model tests, *Proc., Inst. Mech. Engrs*, **168**(20) (1954).
106. Reference 1, Chapter 58.
107. DÈRIAZ, P. E. Water turbines general characteristics, Reference 84, Vol. 2.
108. DOLAND, J. J. *Hydro Power Engineering*, Ronald Press, New York, 1954.

109. HADLEY, B. Governing of water turbines, Reference 84, Vol. 2, 2nd edn, 1970.
109a. LUNDGREN, G. H. and STILLE, C. E. Governing of water turbines, Reference 84, Vol. 2.
110. AGNEW, P. W. and BRYCE, G. W. Optimising turbine operation by electronic governing, *Int. Water Power Dam Constr.*, January 1977.
111. BENKÖ, G. B. Governing turbines for transient loads, *Int. Water Power Dam Constr.*, April 1981.
112. LEVA, DE F. Hydro plant frequency regulation, *Int. Water Power Dam Constr.*, May 1986.
113. Reference 87, Chapter 11.
114. NEUMÜLLER, M. and BERNHAUER, W. Stauregelung und Abflussregelung an Laufwasserkraftwerken (Control of headwater level and discharge at run-of-river plants), *Wasserwirtschaft*, 1969/10.
115. NEUMÜLLER, M. and BERNHAUER, W. Stauregelung und Abflussregelung an Laufwasserkraftwerken mit automatischen Verfahren (Automatic control of headwater level and discharge of run-of-river plants), *Wasserwirtschaft*, 1976/9.
116. Reference 11, Chapter 61.
117. Reference 72, Chapter 9.
118. COTILLON, J. Advantages of bulb units for low-head developments, *Int. Water Power Dam Constr.*, January 1977.
119. MIYAKAWA, S. Characteristics of generators for small-capacity schemes, *Int. Water Power Dam Constr.*, March 1983.
120. FROST, A. C. H. and MORTON, C. H. Switchgear and protection, Reference 84, Vol. 2.
121. Reference 1, Chapter 65.
122. FROST, A. C. H. Transformers, Reference 84, Vol. 2.
123. FROST, A. C. H. Control and communication equipment, Reference 84, Vol. 2.
124. Reference 72, Chapter 8.
125. ZOWSKI, TH. Trashracks and raking equipment, *Int. Water Power Dam Constr.*, October 1960.
126. HERZOG, M. Schwingungen von Rechen im strömenden Wasser (Vibration of Trashracks in flowing water), *Wasserwirtschaft*, 1985/5.
127. Reference 1, Chapter 45.
128. Reference 1, Chapter 52.
129. Reference 72, Chapter 8, Figs 8/4, 8/12 and 8/13.
130. Reference 1, Chapters 68–72.
131. Reference 11, Chapter 68.
132. RHEINGANS, W. J. Power swings in hydroelectric power plants, *Trans. Amer. Soc. Mech. Engrs*, April 1940.
133. SAGAWA, T. Reduction of noise and vibrations in a hydraulic turbine, *Trans. Amer. Soc. Mech. Engrs,* December 1969.
134. LELIAVSKY, S. *Hydro-Electric Engineering for Civil Engineers* (Design Textbooks in Civil Engineering, Vol. 8), Chapman and Hall, London/New York, 1982, Chapter 5.
135. Reference 53, Chapter 7, Sections 7.1 and 7.2.

136. MOSONYI, E. Improvement of hydro-electric schemes by hydromechanical investigations and scale model experiments, Economic Commission of Europe of the UN, symposium, Dubrovnik, Yugoslavia, 1970.
137. HILGENDORF, J. Konstruktive Entwicklung der Rohrturbinen bei der Firma Voith, erläutert an ausgeführten Anlagen (Development of bulb turbines at the firm Voith, demonstrated by erected plants), *J. M. Voith, Forschung und Konstruktion* (Fed. Rep. Germany), 15/1967.
138. SCHEUER, F. Für und wider die Rohrturbine (Pros and cons for the tubular turbine), *Elektrotechnik und Maschinenbau*, 1974/5.
139. SCHMIDT, E. Rohrturbinen im Kraftwerk Ottersheim-Wilhering (Tubular turbines in the O.-W. plant), *Österr. Z. Elektr.*, 1973/10.
140. KHANNA, J. K. and BANSAL, S. C. Cavitation characteristics and setting criteria for bulb turbines, *Int. Water Power Dam Constr.*, May 1979.
141. MEIER, W. and MILLER, H. Die Entwicklung der STRAFLO-Turbine (Development of the STRAFLO turbine), *Bulletin des SEV/VSE*, 17/1978.
142. MILLER, H. *The STRAFLO-Turbine*, printed manuscript, Escher Wyss.
143. Waterpower '79, Int. Conf. Small-Scale Hydropower, Washington DC, October 1979, organized by US Army Corps of Engineers and US Department of Energy.
144. First European Conference on Small Hydro, Monte Carlo, December 1982, organized by the editor of *Int. Water Power Dam Constr.*, papers and proceedings in two volumes.
145. First International Conference on Small Hydro, Singapore, February 1984, organized by the editor of *Int. Water Power Dam Constr.*, papers and proceedings in two volumes.
146. Second International Conference on Small Hydro, Hangzhou, P.R. of China, April 1986, organized by the editor of *Int. Water Power Dam Constr.*, papers and proceedings in press.
147. *Microhydropower Handbook*, US Department of Energy, 1983.
148. Reference 72, Chapter 14.
149. Reference 81, Chapters 110–117.
149a. Hangzhou Regional Centre for Small Hydro Power, *Small Hydro Power in China* (a Survéy), Intermediate Technology Publications, London, 1985.
150. VISCHER, D. Der Einfluss der Wasserkraftnutzung auf die Umwelt (Effect of hydropower utilization on environment), *Österreichische Wasserwirtschaft*, 1975/11–12.
151. SCHIMUNEK, K. *Ökologie und Kraftwerksbau (Ecology and Hydroplant Construction)*, ed. A. F. Koska, Donau Strom, Wien/Berlin, in co-operation with Österreichische Donaukraftwerke, 1984.
152. SCHIECHTL, H. Umweltgestaltung bei Wasserkraftanlagen am Lech (Environmental aspects for water power plants along the river Lech), *Wasserwirtschaft*, 1984/3.
153. Reference 72, Chapter 15.
154. *Environmental, Health and Human Ecologic Considerations in Economic Development Projects*, World Bank, Washington DC, 1974.
155. UNKART, R. *Grosswasserbauten im Lichte von Raumplanung, Naturschutz und Bauordnung (Large-Scale Hydraulic Structures with Respect to Regional Planning, Nature Conservation and Building Code)*, Proc. seminar on

Hydraulic Structures and Environment, ed. O. J. Rescher, Inst. Hydraulic Structures, TU Wien, 1983.

156. LUNA, B. L., CLARKE, F.E. *et al. A Procedure for Evaluating Environmental Impact*, Geological Survey circular No. 645, Washington DC, 1971.

157. BISHOP, A. B. An approach to evaluating environmental, social and economic factors in water resources planning, *Water Resources Bulletin*, **8**(4) (1972).

158. MOSONYI, E. Evaluation of physical resources for small-capacity hydropower plants, Keynote address presented to 2nd Int. Conf. Small Hydro, Hangzhou, P.R. of China, organized by the editor of *Int. Water Power Dam Constr.*, April 1986.

159. CARLISLE, R.K. and LYSTRA, D.W. Environmental impact assessment methodology of small scale hydroelectric projects, Conf. Proc., Waterpower '79, 1st Int. Conf. Small-Scale Hydropower, Washington, DC, US Govt. Printing Office, 1979.

160. MOSONYI, E. and BUCK, W. Die Grundlagen der Wirtschaftlichkeitsanalyse und Kostenverteilung (Fundamentals of economic analysis and cost allocation), *Proc. 4th Extension Course on Hydrology*, Deutscher Verband für Wasserwirtschaft Karlsruhe, 1972.

161. DE GARMO, E. P. *Engineering Economy*, 3rd edn, Macmillan, New York, 1960.

162. SAMUELSON, P. A. *Economics: An Introductory Analysis*, 5th edn, McGraw-Hill, New York, 1962.

163. GRANT, E. L. and GRANT, I. *Principles of Engineering Economy*, 5th edn, Ronald Press, New York, 1970.

164. JAMES, L. D. and LEE, R. R. *Economics of Water Resources Planning*, McGraw-Hill, New York, 1971.

165. Reference 72, Chapter 12.

166. Reference 89, Section 2.2.4.

167. HUNTER, J.K. The cost of hydroelectric power, Reference 84, Vol. 3.

168. MOSONYI, E. *Vizeröhasznositás (Water Power Utilization)*, Vol. 2, university textbook, Budapest (Hungary), 1953.

169. Reference 154, Chapter 8, Section 20.

170. KRUTILLA, J. V. and ECKSTEIN, O. *Multiple Purpose River Development*, Johns Hopkins Press, Baltimore, 1958.

171. McINTOSH, P. T. and LA TOUCHE, M. C. D. Cost sharing in a multipurpose project, *Int. Water Power Dam Constr.*, January 1985.

172. GORDON, J. L. Hydropower cost estimates, *Int. Water Power Dam Constr.*, November 1983.

173. BINGLI, DENG. Small hydro in China: progress and prospects, *Int. Water Power Dam Constr.*, February 1985.

Chapter 2

INTAKE DESIGN FOR ICE CONDITIONS

G. D. ASHTON

US Army Cold Regions Research and Engineering Laboratory, Hanover, New Hampshire 03755, USA

NOTATION

A	area of water surface (m^2)
C	Chezy coefficient ($m^{1/2} s^{-1}$)
C	volumetric frazil concentration (dimensionless)
C_p	specific heat of water ($J\,kg^{-1}\,{}^{\circ}C^{-1}$)
D	flow depth (m)
e	porosity
g	acceleration due to gravity ($m\,s^{-2}$)
h	ice thickness (m)
H	total depth (m)
H_e	depth to top of a submerged intake (m)
H_{ia}	heat transfer coefficient, ice to air ($W\,m^{-2}\,{}^{\circ}C^{-1}$)
H_L	head loss (m)
H_2	distance from bottom of intake to stream bottom (m)
k_i	thermal conductivity of ice ($W\,m^{-1}\,{}^{\circ}C^{-1}$)
r	radius (m)
r_0	radius of bar (m)
t	time (s)
T_a	air temperature (°C)

107

T_f	flow temperature (°C)
T_m	melting temperature of ice (°C)
t_{max}	time to complete blockage (s)
T_s	surface temperature (°C)
T_w	water temperature (°C)
V	mean velocity (m s^{-1})
V_c	critical velocity for ice block underturning (m s^{-1})
V_f	volume of frazil per unit area (m)
ψ	collection efficiency (dimensionless)
ϕ	heat loss rate (W m^{-2})
ϕ_i	heat flux through ice (W m^{-2})
ϕ_{ia}	heat loss from ice to air (W m^{-2})
ϕ_w	heat flux from water to ice (W m^{-2})
λ	heat of fusion of ice (J kg^{-1})
ρ_i	density of ice (kg m^{-3})
ρ_w	density of water (kg m^{-3})

1 INTRODUCTION

The design of water intakes is almost always based on a combined consideration of the water supply needs and the characteristics of the source
water body. The water supply requirements generally consist of a required
withdrawal or passage rate, some measure of reliability, and some measure
of quality ranging from temperature in the case of selective withdrawal
structures to the exclusion of debris or trash. The source water body may
be a lake, a reservoir, the ocean, or a river. It is not surprising therefore
that there is a wide range of types, sizes, arrangements, and concepts for
specific intake designs. Where ice forms in or on the source water body,
consideration must also be given either to preventing its ingestion into the
intake or ingesting it in such a condition that it does not interfere with the
water usage. At first glance one is tempted to treat ice at intakes simply as
a debris problem. This is generally a naive approach because the quantities
of ice are orders of magnitude larger than those of debris, and the nature
of ice, other than its slight buoyancy, is often very different. Because of the
diversity of intakes, the approach taken in this text is to characterize ice
conditions that occur in the vicinity of intakes and then describe the
principles behind various counter measures that have been used both
successfully and unsuccessfully to avoid or mitigate ice problems. Finally

some case studies will help demonstrate the large measure of engineering judgment that must be used to effectively accommodate ice conditions at intakes.

A number of reviews have been published putting the particular ice problems of intakes in the overall context of river and lake ice behavior.[3,5,6,30,41] Others[14,25] surveyed ice problems at intakes in some detail.

2 ICE CONDITIONS IN RIVERS

2.1 Freeze-up

The nature of initial ice formation in rivers depends largely on the flow velocity. At very low velocities a sheet of ice forms over the entire surface, interlocks with the banks, and forms an intact ice cover. At low velocities shore ice also rapidly grows out from the banks and covers the river, again forming an intact ice cover. For such rivers, ice generally does not cause problems at intakes unless the ice is broken up and carried downstream because of higher discharges and hence increased velocities. This condition of a low velocity and an intact ice cover is the aim of some schemes to prevent ice problems at intakes. Unfortunately it is often difficult to attain without expensive impounding structures.

At intermediate velocities the initial ice formation is in the form of thin plates on the surface that move with the flow; this ice is termed skim ice. In this case there is sufficient turbulence to prevent an intact cover from forming but not sufficient to entrain the initial crystals of ice into the flow. The total production of ice over a reach under these conditions may pose a problem at intakes downstream; it must be either passed or accumulated in an ice cover that progresses upstream, forming an initial cover that subsequently thickens by thermal growth.

At some higher velocity turbulence is sufficient to entrain the initial crystals, and a form of ice termed 'frazil' is produced. Frazil ice consists of very small crystals that are suspended in the flow initially throughout the depth but subsequently cluster together, flocculate, and rise to the surface, often in the form of loosely organized 'pans' of frazil. Gosink and Osterkamp[26] termed this regime 'layered flow' of frazil. Since this type of ice causes most of the river ice intake problems, we will discuss it later in more detail. At even higher velocities the turbulent mixing prevents the flocculation and formation of pans, and the frazil is distributed throughout the flow without organized structure of pans on the surface. Figure 1 distinguishes these regimes of initial ice formation on the basis of the

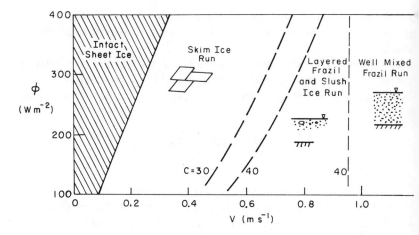

Fig. 1. Regimes of initial formation of ice in rivers (after Matousek[29]).

velocity V and the heat loss rate ϕ. The division between skim ice run and layered frazil runs depends on the roughness coefficient, here characterized by the Chezy coefficient (defined by $V = CR^{1/2}s^{1/2}$ where R is the hydraulic radius); the lines for $C = 30$ and $40\,\text{m}^{1/2}\text{s}^{-1}$ are shown. The vertical line distinguishing layered frazil runs and well mixed runs is based on Gosink and Osterkamp's suggestion that the transition is described by

$$V = C/42 \tag{1}$$

These threshold boundaries should not be taken as exact, but they do correspond reasonably well with field data from various rivers.

Frazil can also form anchor ice, which is the deposit, growth, and accretion of frazil on bed materials. The crystals interlock somewhat and they may grow larger than those of frazil moving with the flow. Anchor ice forms only in areas of open water. Its importance to intakes is that it often rises to the surface during warmer periods. It then appears much like frazil slush masses, with the added danger that it may bring rocks and stones with it.

2.2 Frazil Ice

Frazil ice consists of small particles of ice that form in water that has been supercooled below the melting point. While often associated with rapids in rivers, it also forms in rapidly moving open areas of rivers once the water has been cooled to 0°C, on the surface of lakes and reservoirs when the wind conditions prevent a stable ice cover from remaining intact, and in the ocean in open areas subject to wind mixing. It is predictable in the sense that if the water surface is open to the atmosphere and the rate of cooling through the 0°C point is known, then the production rate can be calculated using a simple energy balance. Since all the heat loss must result in ice production once the water is at 0°C, the rate of frazil production per unit area is given by

$$\frac{\mathrm{d}V_f}{\mathrm{d}t} = \frac{\phi A}{\rho_i \lambda} \tag{2}$$

where V_f is the volume of frazil per unit area, t is time, ϕ is the heat loss to the atmosphere from the water surface, A is the area of open water, ρ_i is the density of ice, and λ is the heat of fusion. Frazil generally forms as tiny discs with diameters typically a millimeter or less and thicknesses 1/5 to 1/10 of the diameter. The initial nucleation is at the surface, but the number of particles multiplies rapidly and they are easily entrained to great depths in river flows. As water containing the frazil particles moves downstream, the particles flocculate together, and the flocs rise to the surface, eventually forming a loose cover consisting of pans or floes that may be characterized as floating slush masses. With further heat loss the surface may freeze, pans of ice are formed and eventually may attain sufficient structural integrity to bridge the river and initiate a stable ice cover. The ice cover on the river then protects the water below from further heat loss to the atmosphere. As a consequence frazil does not form beneath solid ice covers, although it may be transported and deposited at considerable distances beneath the cover.

Frazil ice forms in nearly all rivers. The amount that forms depends on the extent of open water, the rate of heat loss to the atmosphere, and the duration of the heat losses. During the frazil formation period the matrix water is supercooled to between 0°C and -0.05°C; in these conditions the particles grow actively and tend to attach to any cold object they contact. When the water temperature returns to 0°C and the ice and water are in equilibrium with each other, or when the water is slightly above 0°C, the particles no longer grow and the frazil is termed 'passive'. Passive frazil, when dispersed throughout the flow, is generally not a problem to intakes.

A typical time–temperature curve of a body of water undergoing super-cooling and frazil formation is presented in Fig. 2. From A to B the water is cooling at a rate

$$\frac{\mathrm{d}T_w}{\mathrm{d}t} = \frac{\phi}{\rho_w C_p D} \tag{3}$$

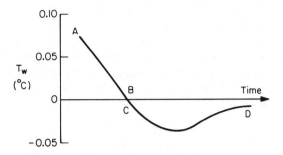

FIG. 2. Water temperature during frazil formation.

where T_w is the water temperature, ρ_w is the water density, C_p is the specific heat, and D is the depth of the flow. As the temperature passes below $0°C$ it supercools, and at point C ice begins to form; the water continues to cool until the ice production is in balance with the heat loss and the temperature gradually returns toward $0°C$ at D. As long as the water continues to lose heat there will be a residual supercooling, ice will continue to be produced in accordance with eqn (2), and the frazil will remain active.

One means of forecasting the appearance of frazil is to monitor the water temperatures upstream of the intake and extrapolate to point B. Since heat losses are generally greatest at night and least during daytime (due to solar radiation), the extrapolation may show that frazil will not appear before daybreak, or at least will not be active. This cooling curve generally occurs synoptically over the entire upstream reach, and passive frazil may reach the intake as a result of production well upstream many hours earlier.

2.3 Thermal Growth of Ice

It is useful to be able to calculate the thickness of an ice sheet exposed to the atmosphere at an air temperature T_a (below $0°C$). Figure 3 shows that the heat loss ϕ_{ia} from the top surface of the ice T_s to the atmosphere at a temperature T_a is calculated using a bulk heat transfer coefficient H_{ia} in the form

$$\phi_{ia} = H_{ia}(T_s - T_a) \qquad (4)$$

The heat loss through the ice sheet ϕ_i is by conduction and is of the form

$$\phi_i = \frac{k_i(T_m - T_s)}{h} \qquad (5)$$

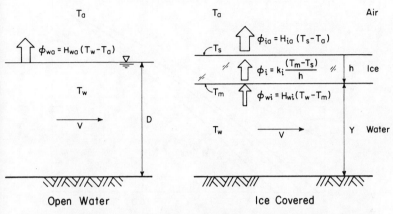

FIG. 3. Definition sketch for heat losses from rivers with ice covers.

where k_i is the thermal conductivity of the ice, T_m is the melting point and the temperature of the ice–water interface, and h is the ice thickness. At the bottom an energy balance gives the thickening rate in the form

$$\rho_i \lambda \frac{dh}{dt} = \phi_i - \phi_w \tag{6}$$

where ϕ_w is the heat flux from the water to the ice. In most cases ϕ_w may be neglected since the water is usually at 0°C. Combining eqns (4), (5) and (6), eliminating T_s, and integrating results in

$$h = \left[\frac{2k_i}{\rho_i \lambda} (T_m - T_a) t + \left(\frac{k_i}{H_{ia}} \right)^2 \right]^{1/2} - \frac{k_i}{H_{ia}} \tag{7}$$

Typically H_{ia} is in the range 10–30 W m^{-2} °C^{-1}, with 20 a good value to use without any other measurements or detailed energy budget calculations. In Fig. 4 eqn (7) is plotted for $H_{ia} = 10, 20$ and 30 W m^{-2} °C^{-1}. The results in Fig. 4, as a function of freezing degree-days are applicable to snow-free ice covers over 0°C water. The initial ice growth is linear in time, and as it gets thicker it grows with $\sqrt{t}$. This figure may be used for times ranging from hours to days or weeks. When used for only part of a day the value of H_{ia} should be estimated using energy budget methods to determine ϕ_{ia}, since the contribution of solar radiation is a gain during daylight but is zero at night. If the solidification of an ice cover initially composed of frazil with porosity e is considered, then λ should be replaced by $e\lambda$ since only the liquid water fraction must be frozen.

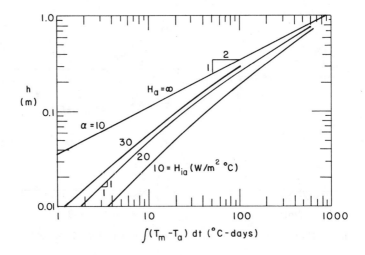

FIG. 4. Ice thickening as a function of accumulated freezing degree-days.

2.4 Accumulation of Ice

When the floating sheet ice or the frazil slush runs arrive at an obstacle, they accumulate into a floating ice cover that progresses upstream. The initial underturning of a block of ice arriving at the ice edge occurs if the velocity is greater than[4]

$$V_c = \left[gh \left(1 - \frac{\rho_i}{\rho_w} \right) \right]^{1/2} \frac{2(1 - (h/D))}{[5 - 3(1 - (h/D))^2]^{1/2}} \tag{8}$$

where g is the acceleration due to gravity. If the velocity is less than this, the individual floes accumulate by juxtaposition in a single layer. If the velocity is greater than this, the ice will accumulate to a more or less uniform thickness, which can be predicted reasonably well although the details are more than can be presented here. For a good discussion of these accumulations, see Beltaos[9,10] and Calkins.[13]

2.5 Break-up

Most of the above discussion has concerned the period of formation of river ice, which may be the entire winter in temperature climates. In the spring or other warming periods break-up occurs. The problems that this causes at intakes are not due to blockage as much as to the high water levels and physical damage associated with ice jams. Ice jams tend to form in the same places along rivers although not necessarily every year. Nevertheless, the regularity of recurrence at the same location forces the designer to explore historic records to determine if ice jams will affect the intake. It is sometimes possible to construct a stage-recurrence interval diagram for ice-jam floods; in general this will differ from the stage-recurrence relationship of non-ice floods and will often show higher stages.[23] Typical ice jams begin where the slope becomes flatter (in the downstream direction); the classic case is the inlet to a reservoir. If the intake alters the geometry of the river, then an assessment must be made as to whether the altered geometry will inhibit or promote ice jams.

One successful means of evaluating these effects is the use of hydraulic models with either real ice which requires refrigerated test rooms, or simulated ice.[34] Considerable success has been achieved using plastic pellets to simulate the ice accumulations with some treatments to eliminate scale effects associated with surface tension.[34] Pariset *et al.*[36] describe such a model test and field validation associated with reclaiming land from the St. Lawrence River at Montreal for the site of the 1967 World Exhibition.

If the change in geometry results in an altered longitudinal depth profile, considerable success has been achieved in predicting the ice accumulation thicknesses and stage heights using analytical models largely based on the initial work of Pariset *et al.*[36] but since refined and validated against field measurements. A recent summary of this technique is given by Beltaos.[9]

If the geometry results in different plan geometries then combinations of model studies and consultation with experienced people is recommended.

The assessment of structural integrity against forces due to ice is generally based on assuring that the structure is stronger than the ice that hits it. In its simplest form this is done by estimating the crushing strength and multiplying by the width of the structure and the thickness of the ice. The subject is beyond the scope of this chapter, and reference is made to reviews of the current state-of-the-art (see e.g. Chapter 3 in reference 5).

3 ICE CONDITIONS IN LAKES

3.1 Static Ice Formation

In relatively small, protected lakes the initial ice cover forms very quickly over the entire surface and, if not broken up as a result of warming and/or wind action, grows steadily in accordance with the analysis of Section 2.3. Snow on top of the ice insulates and slows the growth, but this is often offset because the snow load submerges the ice cover and water infiltrates into the snow and freezes to form snow ice. From the point of view of intakes, lake ice covers are relatively benign. A major design consideration is to avoid placing the intake in the space that the ice normally occupies. If the intake structure is exposed, some consideration should be given to ice loads, particularly due to thermal expansion (see reference 5, Chapter 3). A completely submerged intake in an ice-covered lake is little affected by the ice cover above.

One common misconception is that the water beneath the ice cover is at or near 4°C which is the temperature of fresh water at its maximum density. Seldom is this the case, even at the bottom, simply because wind generally delays the initial ice cover after the isothermal 4°C state is reached, and substantial cooling may occur before the formation of the first complete ice cover, which then effectively shuts off the water beneath from the effects of the atmosphere. Water temperatures of 1–2°C are more common for smaller lakes.

3.2 Dynamic Ice Formation
In larger lakes the initial ice cover forms considerably later than in smaller lakes. On the one hand, the thermal mass of deeper lakes takes longer to cool to 4°C; on the other hand, wind is more effective in destroying any initial ice skim that may form during still, cold periods. With very large lakes a complete ice cover may rarely form during the winter. Rather, there will be shore-associated ice formations of many types and large broken ice fields that move in the lake in response to wind and currents. In open water areas the wind will often prevent thin sheets of ice from forming; instead, large quantities of frazil will be produced in association with wind mixing. The frazil may extend to great depths, and during formation the entire column of water may be supercooled so that the frazil is in the active state. It is the withdrawal of this frazil-laden, supercooled water that causes most lake intake problems.

These dynamic ice formation modes may occur for only a short time at the beginning and end of the ice cover period for intermediate-sized lakes, or they may extend over the entire winter period in larger lakes. The best guide, of course, comes from observations. If there is lack of regularly kept records, local people may be able to provide historical information, particularly those whose activities are affected by the ice, such as boaters, commercial and recreational fishermen, and in many cases, other enterprises or communities who have intakes installed in the same area. The fact that a nearby intake has not experienced ice problems should not be a reason for assuming that the planned intake will not experience ice problems. There are many cases where similar intakes on the same water body have quite different records with regard to ice. Unfortunately our knowledge of ice behavior at intakes is too imprecise to always explain these differences. In such cases a detailed examination of the extent, nature and timing of the ice cover at the sites, of the geometries of the intakes, and the rates of withdrawal may give some clues, especially when related to the ice formation processes. In the case studies presented later, three intakes on the St. Lawrence River at Montreal experiencing quite different ice problems are described and related to the particular ice formations that occur.

3.3 Shoreline Ice Formation

The nature of shoreline ice formation is extremely important to the design of intakes, both for shore intakes where the ice obviously may interfere with the withdrawal and for offshore intakes where the withdrawal piping must pass under or through the ice. Our knowledge of shoreline ice features is a potpourri of observations, mostly qualitative, sometimes documented by photography, and only rarely quantified or subjected to any kind of generalized analysis. Accordingly there is little systematic guidance for assessing the nature of the ice formation to be expected at a new site. Fortunately there seems to be a certain amount of regularity of formational type from year to year at any particular site, so even one year's observation will often identify the most prevalent form of ice to be expected. As with other natural phenomena, however, a short observational period may not identify the extreme events, and these have on occasion resulted in serious shutdowns of water supply. An excellent and enlightening account of an extreme event was presented by Foulds,[21,22] who pointed out that the extreme problems tend to occur when the weather is so bad that visual evidence is meager.

There has been one study (N. Gruber, personal communication) that attempted to categorize shoreline ice formation on a large lake and provide a scheme that would permit, together with a long-term meteorological record, a hazard assessment of the various formation types. The study was specific to one site on Lake Ontario, so the criteria outlined in Table 1 are not directly transferable to other sites. Its success against some 20 observations, however, suggests that it is a starting point for other site assessments.

Primary ice formations (Table 1) are divided into smooth ice sheet formations associated with below-freezing conditions and low wind speeds. Two kinds of frazil ice formations are distinguished: a daytime frazil formation for which the air temperature must be cold, the wind low

TABLE 1

SHORELINE ICE FORMATION CRITERIA (FROM ASHTON[3] AFTER GRUBER)

	Prior formation required	Meteorological parameters			
		Air temp. (°C)	Wind speed (m s⁻¹)	Wind direction	Cloud cover
Primary ice formations					
1. Smooth sheet	None	$\leqslant 0$	$\leqslant 5$	–	–
2. Frazil ice (daytime)	None	$\leqslant -1$	$\leqslant 5$	–	> 0.6
3. Frazil icc (night)	None	$\leqslant 0.5$	–	–	< 0.5
4. Slush balls	None	$\leqslant 0$	> 5	–	–
Secondary ice formations					
5. Conglomerate sheet	6, 7, 4 or 8	$\leqslant 0$	$\leqslant 5$	Onshore	–
6. Broken sheet and	1^a or 5^a	$\geqslant 0$	–	–	–
pack ice	1^a or 5^a	–	$\geqslant 3.6$	Onshore	–
7. Frazil slush	2 or 3	$\leqslant 0$	$\leqslant 5$	Onshore	–
8. Pancake ice	(7 or snow) and 1	$\leqslant 0$	$\leqslant 5$	–	–
9. Rafting	6^a	–	> 5	Onshore	–
10. Piling	6^a	$\geqslant 0$	$\geqslant 5$	Onshore	–
11. Ridging	1^a or 5^a	–	> 5	Onshore	–

^{a} Within 24 hour period.

to moderate, and the cloud cover sufficient to block much of the incoming solar radiation; and a night frazil formation for which low air temperatures and little cloud cover are required, the latter to allow significant long-wave radiation loss. If the air temperature is low and the wind is high, the frazil will form a field of slush balls covering the surface.

Secondary ice formations (Table 1) are those that form after the primary ice formations. These also often depend on the wind direction being either onshore or offshore. Thus, a conglomerate sheet may result when a previous formation of slush balls formed under high wind conditions is subsequently blown onshore by moderate winds. Pancake ice, itself a secondary formation resulting from frazil, can be formed into frazil slush and then into conglomerate ice under the influence of onshore winds. A smooth sheet of ice may subsequently form a pack ice cover under offshore winds, only to form a conglomerate sheet when the wind turns to the onshore direction.

The most serious non-frazil shoreline ice conditions are rafting, piling, or ridging, all of which result from previous ice formations being driven by high onshore winds. Rafting is here defined as the movement of one ice cover over or under another, and it generally requires open water between the two covers prior to rafting so that enough momentum can develop. Piling is a more random accumulation that also results from impact of a moving ice sheet or field with the shore or a shorefast ice cover. It can occur very quickly. Piles as high as 4–5 m are not rare. Ridging results from the relative motion of two ice sheets and may be the outcome of either compression or shearing. Wave spray can also sometimes build up very large accumulations at the shore, often extending tens of meters into the lake. In general, frazil causes most of the intake interruption due to flow blockage, while piling and ridging cause most of the interruptions due to damage.

3.4 Ridging and Scouring

Until recently the vertical extent of ridging in large lakes was little appreciated. In Lake Erie parent ice that was only 0·3–0·6 m thick has yielded ridges of sufficient depth to scour the bottom in water 16 m deep. The ridging occurred as the result of large fields of ice moving relative to each other. At their intersection the ice may pile both upward and downward into large ridges. These ridges move under the influence of water currents and wind and have sufficient momentum to gouge or scour the bottom. An examination of the bottom over several years in depths of about 9 m showed not only that there were many scour tracks but also that ridging and scouring occurred several times each winter. The implications for intake design are twofold: first, if the intake is exposed at depths subject to scouring, it must have sufficient structural integrity to survive the ridge impact, and second, the conduit must be protected either by providing structural integrity or by burying beneath expected scour depths. It is common practice to lay the conduit in a trench and provide a protective rock cover over the width of the trench.[39]

There are several means of assessing the hazard of scouring. Observation of ridging in the vicinity suggest that the hazard may exist at depth. If a ridge or pile is stationary while the surrounding ice field moves, it is grounded and probably scouring to some extent. Finally, examination of the bottom will often detect scour tracks. In some cases, as in Lake Erie, sediment transport fills the scours over the course of the summer, so the examination should be conducted as early in the spring as possible.

4 OCEAN COASTAL ICE CONDITIONS

Ice conditions in the ocean along coasts are similar in many respects to those on the coasts of very large lakes. However, saline water, wave action, and especially tidal action tend to make the ice formations more complex and often more severe. Field observations at prospective sites are necessary to identify the types of ice formation to be expected. Examples of the considerations of design with respect to ice are given by Wu,[44] Colonell and Lifton,[16] Cox and Machemehl[17,18] for a high Arctic intake, and by Kuzovlev[27] for an intake on the Caspian Sea.

5 BEHAVIOR OF ICE AT INTAKES

Up to this point, the emphasis has been on describing the nature, kinds, and principles of formation of ice in rivers and lakes that an intake can be expected to experience. The withdrawal of water further affects that ice. In this section the behavior of ice at intakes is described based on both observations and experience at actual intakes as well as some limited laboratory studies.

5.1 Frazil Ice at Intakes

There are several ways in which frazil ice blocks intakes. The frazil can build up directly on the trashrack bars, either by particles directly freezing to the bar elements and subsequently to the ice or by the growth of frazil crystals directly attached to the bars or sides of openings. Intakes can simply be clogged due to large masses of frazil impinging against the opening, sometimes with a subsequent compression of the initially loose mass of frazil floes. Frazil can also accumulate over the flow passage to the point where the flow opening is greatly constricted. The first mode of accumulation on individual bar elements we will term 'icing', the second mode 'clogging', and the third mode 'accumulation'. There are many reports describing intake blockage by frazil; however, it is often difficult to determine the mode of the blockage from the usually incomplete descriptions. Direct observations of the nature of the ice and its location are occasionally reported; more often the evidence is little more than a statement that the flow became constricted and pumping capacity decreased. There have been few fundamental studies of the frazil blockage phenomena, and only recently have controlled laboratory investigations begun to provide a basis for rational design. Thus what follows is to a considerable degree based on inference rather than on a large body of fundamental results.

When water containing frazil is supercooled, the frazil particles are in an 'active' state; that is, they are actively growing and will attach themselves to cold objects that they contact. In the initial stage of icing of, say, a bar element of a trashrack, frazil particles will adhere directly to the surface of the bar. The particles seem initially to attach more easily to surfaces with a high thermal conductivity, such as metals, than to surfaces with a low thermal conductivity, such as plastics or wood. The general view is that the substrate, having been cooled below 0°C by the passing water, acts as a sink for removing the latent heat associated with the bonding of the particle to the substrate. Williams[43] pointed out that the applicable property of the material is the 'conductivity capacity' $\rho C_p \sqrt{k}$, where ρ is the material density, C_p is the specific heat, and k is the thermal conductivity. Williams[43] found that there was less tendency for frazil to cling to plastic bars than to steel; however, once an initial ice buildup has occurred, of course, the subsequently arriving ice 'sees' ice, regardless of the underlying substrate. Thus, the use of plastic, plastic coatings, wood, or other similar materials may not cure a frazil adhesion problem, but they may help and their use is encouraged.

There seem to be two modes by which the ice builds up on trashrack bars. At one extreme the buildup is dominated by deposition and attachment of the particles carried by the flow to the bar element. On the other extreme the buildup is caused by the growth of the individual particles into fairly large, thin ice crystals, sometimes as large as a few centimeters. This crystal growth is driven by 'irrigation' of the crystal edge by supercooled water, which supplies the heat sink for the heat transfer from the crystal to the water. Using edge growth heat transfer relations (see e.g. Daly[20]) it may be shown that the edge growth rate is proportional to $(V/r)^{1/2}(T_m - T_f)$, where V is the velocity by the crystal, r is the radius of the edge of the crystal (half the thickness), T_m is 0°C and T_f is the temperature of the flow. Using the appropriate constants and coefficients and supercooling of 0·04°C yields edge growth rates of up to about a centimeter per hour for 1 mm thick crystals. This kind of growth is probably the explanation for 'glass-like plates' of ice reported in some observations of frazil blockage.[8] A similar crystal growth habit sometimes occurs in anchor ice deposits on rocks.

In the depositional mode of frazil buildup the supply of crystals from the flow determines the rate of buildup on the bar element. The ice crystals freeze to the bar elements (and then to the ice) only when the flow is

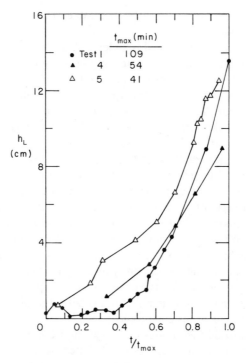

FIG. 5. Head loss across a trashrack while building up with frazil (after Daly[19]).

supercooled. The rate of buildup on an individual bar is proportional to the concentration of the frazil in the flow. Daly[19] recently analyzed the buildup of ice on a submerged bar with radius r_0 and spacing s. The volumetric rate of buildup per unit length of the bar is expressed as

$$\frac{d(\text{Vol})}{dt} = 2\pi r \frac{dr}{dt} = \psi r V C \frac{\rho_w}{\rho_i} \tag{9}$$

where ψ is a 'collection efficiency', r is the radius of the buildup, V is the mean velocity through the opening, C is the concentration of frazil in the flow, and ρ_w and ρ_i are the densities of water and ice. The time to complete occlusion is then given by

$$t_{max} = \frac{(s - r_0)\pi\rho_i}{\psi V C \rho_w} \tag{10}$$

Larger spacings, smaller velocities, and smaller concentrations all result in longer times to occlusion. Figure 5 shows the results of several experiments in a flume in which the head loss h_L was measured as frazil built up on a rack of 1·25 cm bars at 4 cm spacings. The time to occlusion ranged from 40 to 110 min. However, the increase in loss occurred slowly at first and then rapidly to complete closure. The message for intake operators is clearly that frazil will rapidly cut down the flow once it begins building upon intake racks. The same analysis described above results in a time dependence of the head loss of the form

$$h_L \alpha \frac{1}{1 - (t/t_{max})} \tag{11}$$

Thus, while the head loss increases very rapidly just before occlusion it is not instantaneous as sometimes reported by operators, although the time over which the head loss is significant may be quite short.

Trashracks may also be blocked by large masses of frazil impinging against the trashracks. Generally these large masses of frazil are seen at the surface of the upstream flow, and the initial clogging is at the top section of the trashrack. Once that region is blocked, the next frazil is easily drawn down and the clogging progresses downward.

Although ice accumulation on trashracks is the mode most often blamed for intake blockage, it is not always the culprit. Sometimes frazil deposits form in the intake chamber or intake conduit and significantly constrict the flow. Whether and where this has occurred in a specific instance is often difficult to determine without visual inspection. Sometimes these frazil clusters will later detach, and pieces of the accumulation arriving downstream (or in the intake region when backflushed) will have shapes suggesting where they were attached.

5.2 Broken Ice at Intakes

There are occasions when the ice that arrives at intakes is broken up into slabs and fragments and it is desirable not to pass them into the intakes. Little is known about the depth below the surface that an intake must be placed to avoid entrainment of fragmented ice. The depth must be at least greater than the thickness of the cover that will accumulate under the action of an approach velocity V (Fig. 6). Using the results of Pariset and Hausser,[35,36] that depth is given implicitly by

$$\frac{V}{[gH_e(1 - (\rho_i/\rho_w))]^{1/2}} = 2^{1/2}\left(1 - \frac{H_e}{H}\right) \tag{12}$$

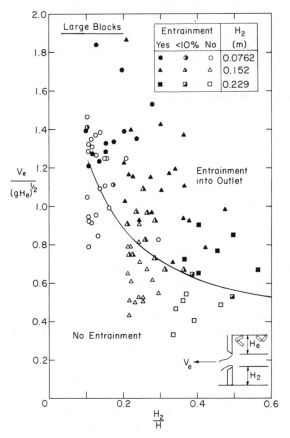

FIG. 6. Entrainment of ice blocks into a submerged outlet; experimental results.[42]

However, Stewart and Ashton[42] found in model tests that eqn (12) was not an adequate criterion since, even at depths greater than that indicated, significant numbers of fragments were entrained into the outlet. They proposed an equation that discriminates between entrainment and non-entrainment events of the form

$$\frac{V_e}{(gH_e)^{1/2}} = 0{\cdot}4\left(\frac{H}{H_2}\right)^{1/2} \tag{13}$$

Their data and the discrimination equation (13) are shown in Fig. 6 and provide limited guidance.

6 COUNTER MEASURES

There are a wide variety of measures that can be taken to counter the adverse effects of ice at intakes. Just as the nature of the ice problems at intakes is highly site-dependent, so too are the means of avoiding or mitigating those problems. Ideally the potential ice problems are considered in the design stages and the design is adapted to the expected ice conditions. More often, however, the ice problems are discovered after construction is complete or there are inherent constraints on the design that force the ice problems to be dealt with by modifying the original design or by changing the operation of the intake.

6.1 Changing the Ice Regime

Most of the ice problems at intakes are due to accumulation, ingestion, or blockage by small pieces of ice, ranging from frazil to the fragmented ice often called 'brash'. Many of these problems would be alleviated if the ice regime were changed from one that produced frazil and brash to a regime in which a stable, intact sheet of ice forms over the reaches upstream of the intake, and the water flows free of ice under this cover to the intake.

There are several purposes that a stable ice cover upstream of an intake may serve. If a significant portion of the reaches that contribute frazil or brash ice can be covered with a stable ice sheet, then the production will be reduced and the problem essentially avoided. Where a stable ice cover cannot be induced over the entire contributing reach, it is sometimes induced over a limited reach just upstream to provide a storage zone for arriving ice or to change the frazil from its active to its passive state. In rare cases it is possible to divert the ice away from the intake.

The main ways of changing the ice regime are to reduce the velocity or to stabilize the cover. By the principle of continuity, the velocity can be reduced either by increasing the flow area or by reducing the discharge. The flow area can be reduced either by excavating a larger cross section or by raising the water level. Figure 1 shows that the velocity must be reduced at least to $0.7\,\mathrm{m\,s^{-1}}$, and even then some stabilization of the cover is required either by letting the ice cover progress upstream from the intake or by installing floating booms to initiate the ice cover. If the velocity can be reduced below about $0.3\,\mathrm{m\,s^{-1}}$, the ice cover will generally form an intact sheet soon after initial formation with little, if any, assistance. The advantage of decreasing the velocity by increasing the flow area is that it is a permanent change and thus does not alter ordinary operational procedures at the intake.

When sufficient flow control is available, the discharge can be reduced temporarily until a stable ice cover has formed. This ice cover can then be subjected to higher velocities without disruption, and higher flows can be maintained throughout the remainder of the cold period. This technique has been used at many sites, but there have been few attempts to generalize the experience. A related problem is determining how much variation in discharge can be allowed without disrupting the stable ice cover.

The general procedure for establishing the ice cover is as follows. When the water has cooled to 0°C and ice production periods are forecast, the discharge is reduced so that sheet ice rather than frazil is formed. The ice cover is allowed to stabilize and thicken by freezing. After several days to a week, the discharge is gradually increased. It is not known how high a discharge can be tolerated without disrupting the cover. Similarly we do not know how much variation in discharge with its accompanying long waves and slope changes can be tolerated. Nevertheless this general technique has been used in many instances. Examples are described by Billfalk[11] for a river in Sweden and by Boulanger et al.[12] for the Beauharnois Canal in Canada. Besides avoiding the ice production associated with open water, this method also reduces the heat loss along the covered reach compared to that along a cover formed with much shoving and thickening.

The length of ice cover necessary to change active frazil to passive largely depends on how long it takes for the supercooled ice–water mixture to go from the maximum supercooling to the residual supercooled state (point D in Fig. 2) plus some time for the residual supercooling to be removed by the increase in temperature due to viscous dissipation in the flow. Laboratory measurements by Carstens[15] suggest that this time is on the order of 10 min which, for a flow velocity of $0.5 \, \mathrm{m \, s^{-1}}$, translates into a cover length on the order of 300 m. This cover length can only affect the properties of the frazil, in particular the 'stickiness', and the frazil may be transported many tens of kilometers beneath ice covers if there is no intervening slow velocity zone where it can settle out. Nevertheless the change from the active to the passive state can often greatly alleviate intake frazil accretion and blockage.

Often artificial stabilization can induce an intact, stable ice cover on a flow where the ice is otherwise continuously moving. The most common technique is to place ice booms either completely or partially across the flow surface. Typically ice booms are constructed of timber elements tied together with cables and anchored to the bed or shore. They work only when the velocity is about $0.7 \, \mathrm{m \, s^{-1}}$ or less (some have occasionally

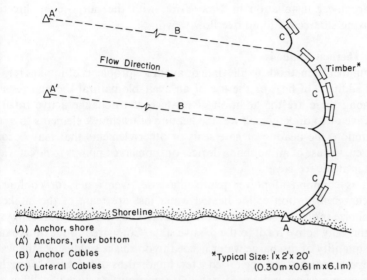

FIG. 7. General ice boom features (from Perham[38]).

performed adequately at slightly higher velocities). They are almost always used to initiate an ice cover, which subsequently thickens and freezes to form the stable cover, and their overall purpose is either to prevent fragmented ice from being carried to the intake or to prevent frazil production by covering the open water. Perham[38] described representative ice boom details along with other ice retention structures. Figure 7 shows general ice boom features.

In some cases piling dolphins upstream and just offshore of an intake have been used to protect an intake from moving ice. Typically the protected downstream zone is five to ten times the distance that the dolphin is placed offshore.

In some cases it has been possible or desirable to divert the ice away from the intake, thus avoiding the difficulties of dealing with the ice at the intake. Perhaps the best known ice diversion is at the Burfell hydroelectric complex in Iceland. There, ice arriving at the dam is diverted from the main flow via an ice skimming weir and is passed around the main intakes in an ice sluice.[40] Morozov[31] described an ice deflecting structure or 'floating wall' that diverts floating ice past the intakes of a pumping station intake on the Kondoma River. And finally, to show the full range of methods that can be used to divert ice, a submerged motor-driven propeller has been used to create a surface flow away from an intake of a small hydropower installation in New York, with the purpose of directing arriving surface ice to an overflow weir.

6.2 Thermal Methods

A number of methods to alleviate or avoid ice problems of intakes rely on the addition of heat or the use of an available natural thermal reserve. Among these are the addition of waste heat to change active frazil to passive or to directly melt ice, the heating of trashrack elements to avoid accretion, the heating of gate seals or other elements that may become frozen, the use of air bubbling devices or submerged pumps to direct warm water at the ice cover.

It is common at power plants that use river water for cooling to discharge a portion of the heated water just upstream of the intake to minimize problems with frazil at the intake. The effect is almost always to change any active frazil to the passive state. Compared to the intake rate, the quantity of discharge water needed to do this is small since the water temperature must only be raised a few hundredths of a degree. More than this amount is generally used, and this is appropriate since it is difficult to achieve complete mixing. Occasionally the geometry of the intakes and outfalls is such that it is possible during ice periods to short-circuit the outflow partially back to the inflow, significantly raising the intake water temperatures. This generally causes little loss in cooling efficiency since the pumps are ordinarily sized for cooling at the warmer summer temperatures.

Similarly, in rare cases it is possible to site an intake downstream of another discharge of waste heat so that the water is withdrawn from the warm plume of the waste heat discharge.

The most common use of heat to alleviate ice problems at intakes is the practice of heating the trashrack elements, thereby preventing the arriving frazil from adhering to the bars. This method is very successful where the problem is due to buildup of ice on the bars; it is less successful and sometimes ineffective if the problem is due to blockage by brash ice. Two main elements of cost are associated with trashrack heating: the operational cost of the heat energy and the capital cost of providing the heating apparatus and controls.

Logan[28] reviewed the practice of heating trashracks in some detail and calculated energy requirements based on maintaining a bar element a fraction of a degree above 0°C (somewhat arbitrarily taken as 0·2°C). The resulting energy requirement per area of opening depends on many factors, such as intake velocity, bar spacing and shape, but the resulting

alues were on the order of $500 \, \text{W m}^{-2}$ of opening. Actual field installa-
ions use values on the order of $2000-8000 \, \text{W m}^{-2}$. Sometimes the require-
nents are stated in terms of power per unit volume of water and typically
ave been on the order of $5000-15\,000 \, \text{W m}^{-3} \text{s}^{-1}$. Such a power input will
aise the bulk water temperature only $0.001-0.002°C$, which is not ordi-
arily enough to change the active frazil-laden flow to the passive state
ince measured supercoolings have typically been $0.01-0.04°C$.

The actual delivery of energy to the trashrack elements is almost always
one by resistance heating of the rack elements. Induction heating has
een successfully used in the USSR. Steam or hot water has sometimes
een passed through the bars, but this is generally rejected in favor of
lectrical heating because of thermal efficiency considerations and the
laborate distribution systems required as well as the increased year-round
ead loss associated with larger bar cross sections. Gevay and Erith[24]
lescribed in detail the application of resistance heating to intakes of a
ydropower installation in Labrador, including the design and selection of
he electrical elements such as transformers, buswork, control systems,
nd grounding, and the electrical characteristics of the trashrack itself.

When heat is applied to trashracks it is important to do so before frazil
uns. Once the trashrack is clogged with frazil, turning on the heat will
ave much less effect since the heat is not sufficient to melt much ice.
Conversely there is no need to operate the heating system when frazil is not
oresent in the flow, so the total power consumption through a winter
season is small.

Another common winter problem at intakes that can often be solved
with the addition of heat is the freezing together of parts that must move
relative to each other. Thus gate seals, guide slots, and other elements are
often provided with electrical heating, or piping to deliver hot water or
steam, to prevent the freezing. When included as part of the original
construction, these measures are relatively inexpensive since they can be
placed in the concrete of the structure. When added they may be as
effective but are much more vulnerable to damage and require more
maintenance. In some cases radiant heaters have been directed at problem
areas with some success.

Air bubblers are often used upstream of intakes from reservoirs to keep
the intake area ice-free. They raise warm water from depth and direct that
warm water against the ice cover to induce melting (or prevent freezing).
The heat flux to the surface on which the flow impinges is more or less
proportional to the product of the velocity and the temperature (above
freezing) of the water, so it is essential that the water at depth be above

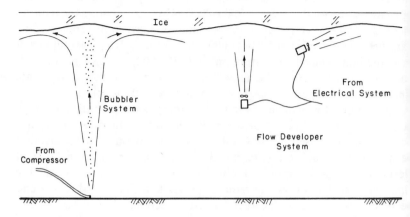

FIG. 8. Schematic of bubbler and flow developer systems.

freezing. Often, however, temperatures as little as 0·1 or 0·2°C will be effective. Ashton[7] described methods of calculating the performance of bubblers to melt ice if the water temperature is known. Figure 8 shows a sketch of a bubbler system and flow developers, described below.

A technique similar to bubblers is to use submerged motor-driven propellers to direct a flow of warm water at the ice cover or upstream side of gates to prevent or alleviate freezing. Ashton[2] compared the relative effectiveness of these pumps with bubblers and concluded that, in terms of power consumption, the two techniques were equivalent. Thus the choice of a bubbler or pump should be made on the basis of other considerations.

It is interesting to compare actual practice, however, in using pumps and bubblers. Most flow developers on the market are of the order of 200–800 W per unit. At equivalent installations most bubbler systems have an effective power much less than that, generally on the order of 25–100 W or less per orifice location. While the comparison is not exact since flow developers are often used at angles near horizontal thus acting to suppress ice as a row of orifices of a bubbler system, there seems to be a tendency to overdesign flow developers or underdesign bubbler systems.

Occasionally pumps have maintained a small open-water area upstream of gates when there was no apparent thermal reserve to draw upon. Apparently the newly arrived water does not have time to freeze into a solid sheet. It is known that operation at 0°C after the ice has formed does not melt the ice.

6.3 Mechanical Methods

A variety of mechanical methods have been used to mitigate ice problems at intakes, including blasting, pneumatic devices, and mechanical removal. While generally not desirable, these brute force methods are often the only measures readily available.

Blasting has sometimes been used to disrupt the ice mass clogging an intake and occasionally has been successful. Baylis and Gerstein[8] described the use of frequent dynamiting of ice in front of lake intake ports in an effort to maintain flow. Dynamite has also been used in an attempt to dislodge frazil accumulated on trashracks. Mussalli[32] described a similar technique using pneumatic guns contemplated for an intake on the Hudson River in New York with the motivation of avoiding some of the problems of blasting. At best, blasting is a hazardous operation with uncertain results and rarely useful until after the problem has developed.

Perhaps the most common method of alleviating ice problems of intakes is mechanical removal. Sometimes large numbers of laborers with long poles attempt to free the ice or rake it off intake grates. Other methods include using materials-handling equipment, such as backhoes and clamshell buckets, or specially designed motor-driven rakes to remove the ice. Even manual chipping at the ice with axes has been used. Besides the hazard to workers, there is often damage to the installation by these methods.

6.4 Hydraulic Methods

Various hydraulic methods have been used in attempts to prevent or alleviate ice problems at intakes. Giffen[25] reported a practice of throttling down the intake flow through the intake that sometimes erodes the ice blockage at crib-intake ports. This procedure was compared favorably to the practice of backflushing, which sometimes did not free the ice while wasting more water than would have been deferred by throttling. Alekseenko[1] reported a more extreme version of throttling. The penstock with a blocked intake had its flow increased by opening the guide vanes, which were then abruptly closed. The 'water hammer' resulted in the trashrack being ingested into the penstock. Nevertheless, when flow interruptions are tolerable and the supply of water for backflushing is available, backflushing may allow intermittent intake operation. A technique using high-air-discharge bubblers or high-velocity water jets has sometimes been successful in 'blowing' accumulated frazil off trashracks.

All of these methods depend on very close monitoring of the ice problems, and successful techniques generally only result from consider-able experience at each installation.

Many of the problems of ice at intakes occur at the trashracks, which are particularly susceptible to clogging and blockage. Often the ice causing the blockage is harmless except for this blockage (although it depends on the receiving units), and removal of the trashracks will often remove the ice blockage problem. It is a fairly widespread practice, but certainly not universal, to remove all or portions of the trashracks during the period of ice problems. This causes a dilemma for the operator, since the trashracks are designed to exclude debris that would cause damage. Since much of the debris that arrives at an intake is waterlogged and submerged, visual observation may not detect the hazard. The decision thus comes to one of balancing two risks: flow blockage by ice and possible ingestion of damaging debris.

6.5 Monitoring Ice Conditions

There are several reasons for monitoring ice conditions at intakes. First, the nature of the problems caused by ice depends on the type of ice involved. Second, while many flow blockages often seem sudden, they almost always develop over a period of time, albeit often short, and monitoring will often allow detection and mitigation before the problem becomes catastrophic. Finally, the cause of many ice problems is not always self-evident, and later diagnosis is helped considerably by having good records of the observed ice conditions, water temperatures, weather conditions (particularly air temperature and wind velocity and direction), head differentials across the entrance and along the intake piping, and flow rates. Most of these observations can be implemented easily by the operators. Recognizing the ice conditions requires a little background reading, both to learn the jargon of ice engineering and to improve the understanding of ice behavior. Water temperature measurements are needed, however, at an accuracy and resolution that is not ordinary, although certainly not difficult. It is routinely possible to achieve ac-curacies approaching 0·01°C using available thermistors. Typically the thermistor itself is enclosed in a small probe that is installed upstream of the intake where it will measure the temperature of the entering water. The wiring that leads to the recorder or readout device is protected by a pipe. When periods of ice formation are expected, the temperature should be monitored closely. The important time, of course, is when the temper-ature approaches 0°C, since that is when ice forms. Often the rate of cooling can be extrapolated to predict when this will occur. In rivers, for example, active frazil ordinarily does not occur during daylight hours except under extreme wind and low air temperatures since the incoming solar radiation offsets the convective cooling, and if the temperature does not decrease to 0°C by then, active frazil will probably not occur. Nelson[33] provided an excellent description of the use of such monitoring to guide the operation of a lake intake located 1·5 km offshore.

Another very simple monitoring technique is the practice of suspending a chain in the intake region and noting whether or not ice is adhering to it. Active frazil will accumulate rapidly in spongy-appearing masses, thus indicating its presence and warning of, or at least identifying, the problem. Baylis and Gerstein[8] show pictures of such a 'tell-tale' chain.

7 CASE STUDIES

There are many descriptions in the literature of ice problems at intakes but few complete descriptions of the details of the problems or the measures taken to mitigate or prevent them from recurring. Very short descriptions of some particular cases are described here, mostly for the purpose of illustrating the diversity of the problems. The sources are referenced for a more complete treatment which includes the detection of the problem, the diagnosis, and the remedial measures.

7.1 River Intakes

Parkinson[37] described the ice problems at three intakes on the St. Lawrence River, all in the vicinity of Montreal. The description is particularly useful since it contrasts the problems, the causes, and the ice conditions at intakes very near to each other.

The Pointe Claire intake withdraws water from the side at a portion of the river covered with a stable ice cover most of the winter. Nevertheless, during early winter before the ice cover forms, active frazil is ingested and collects just inside the intake structure (but not at the trashracks as first suspected). The remedy consisted of providing a traverse-mounted high-pressure sewer cleaning head that is moved back and forth through the accumulated frazil until the obstruction is removed.

The Montreal intake withdraws water from mid-river via an intake structure aligned parallel to the flow, with multiple ports along the two sides. No screens are at the ports, and the frazil blockage was found to occur on the inside walls of the 2·1 m diameter suction lines that lead from the structure 610 m to the shore. Here the solution was to supply warm water in insulated lines from the shore to the outer ends of the suction lines and inject it into the lines by circumferential distribution rings. The warm water is derived from six natural gas burners each with a capacity of 3000 kW. Head loss is monitored and the system operated as required. The operation varies from none in some winters to sporadically over a period of as long as 5 weeks.

The Longueil intake is located near the shore but downstream of a rapidly flowing section of the river. While shore ice sometimes protects it, at other times loose ice and frazil obstruct the intake. Here the solution was to provide an additional intake in the slower-flowing Seaway channel adjacent to the river. Since this channel has a solid ice cover, no frazil is ingested. This emergency intake is used as much as 25 days in a winter season.

7.2 Lake Intakes

Baylis and Gerstein[8] described the intake works and measures taken to circumvent ice problems at a Chicago filtration plant receiving water from Lake Michigan. Both a shore intake and a crib intake 3 km from shore were used (see Fig. 9). At times the shore intake ingested active frazil that clogged within the pump, and a variety of measures were taken to restore pumping capacity, ranging from backflushing to addition of steam. When the offshore crib intake became available, it was used during periods of active frazil, although it was sometimes subject to clogging even with the

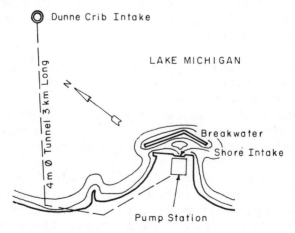

FIG. 9. General arrangement of Chicago intake.

screens removed and with periodic use of dynamite in attempts to clear the openings.

Richardson[39] described a number of intakes and associated ice problems. Among other recommendations is the use of a 'hydraulically balanced' intake configuration, i.e. one in which all the ports have the same entrance velocities. This concept was developed in response to the poor performance of intakes in which the ports with the highest velocities clogged first, progressively shifting the problem to the other ports.

Foulds[21,22] provided excellent descriptions of some particularly difficult ice situations experienced at intakes on the Great Lakes and recommended that lake intakes be located beyond the limits of shore ice, that inlet openings be large enough to prevent frazil from bridging across them (he suggested a minimum dimension of 0·60 m), and that heating be applied to the intake flow before it reaches the intake racks.

There are numerous other accounts of ice problems at intakes in the literature but most are fragmentary at best.

8 CONCLUSIONS

The design of intakes for ice conditions involves assessing the types of ice that will be encountered and either designing to avoid them or providing counter measures to mitigate the blockage. In rivers the type of ice that occurs once the water has cooled to 0°C depends mainly on the velocity and the heat loss rate to the atmosphere. At low velocities sheet ice forms and if stabilized in place it causes little problem. At higher velocities frazil ice forms and has particularly troublesome properties, especially if it is in the active state.

In lakes the problems are largely due to frazil buildup at the intake or supply conduits. In some cases extreme shoreline formations occur and may be either grounded or of such thickness that the bed is scoured by moving ridges.

Monitoring water temperatures and ice conditions allows more efficient operation of intakes to avoid or mitigate ice problems, and in the case of serious stoppages it is a great aid in accurately diagnosing the particular cause.

While ice problems at intakes have caused serious problems for many decades, it is surprising how little quantitative knowledge is available. Only recently has there begun systematic study of such phenomena as the rate of buildup on trashracks as a function of spacing of bars, intake velocities, and concentrations of frazil in the flow. One of the difficulties facing the profession is that there is a great variety of ice conditions and a great variety of intake types and sites. As our experience grows and is documented, improved designs and counter measures will emerge and be based on systematic principles rather than the current individual designer's insight, experience, and ingenuity.

REFERENCES

1. ALEKSEENKO, I. E. Ice difficulties at the V. I. Lenin Dniepr hydroelectric station, *Hydrotechnical Construction,* January 1971, pp. 44–8.
2. ASHTON, G. D. Bubblers and pumps for melting ice, *Proc. IAHR Ice Symp. 1986,* Iowa City, Iowa, Vol. II, 1986, pp. 223–34.
3. ASHTON, G. D. Freshwater ice growth, motion, and decay. In: *Dynamics of Snow and Ice Masses,* ed. S. C. Colbeck, Academic Press, New York, 1980, pp. 261–304.
4. ASHTON, G. D. Froude criterion for ice block stability, *J. Glaciology,* **13** (1974), 307–13.
5. ASHTON, G. D. (ed.), *River and Lake Ice Engineering,* Water Resources Publications, Littleton, Colorado, 1986, 504 pp.
6. ASHTON, G. D. River ice, *Ann. Rev. Fluid Mechanics,* **10** (1978), 369–92.
7. ASHTON, G. D. Theory of thermal control and prevention of ice in rivers and lakes. In: *Advances in Hydroscience,* Vol. 13, Academic Press, New York, 1982, pp. 131–85.
8. BAYLIS, J. A. and GERSTEIN, H. H. Fighting frazil ice at a waterworks, *Engineering News-Record,* 15 April, 1948, pp. 80–3.
9. BELTAOS, S. River ice jams: theory, case studies, and applications, *J. Hydraulic Engng,* **109**(10) (1983), 1338–1359.
10. BELTAOS, S. A conceptual model of river ice breakup, *Can. J. Civil Engng,* **11**(3) (1984), 516–29.
11. BILLFALK, L. Strategic hydropower operation at freeze-up reduces ice jamming, *Proc. IAHR Ice Symp. 1984,* Hamburg, Vol. I, 1984, pp. 265–75.
12. BOULANGER, F., DUMALO, E., LE VAN, D. and RACICOT, L. Ice control study of Lake St. Francis-Beauharnois Canal, Quebec, Canada, *Proc. 3rd Int. Symp. Ice Problems,* Hanover, New Hampshire, USA, 1975, pp. 39–48.
13. CALKINS, D. J. Ice Jams in shallow waters with floodplain flow, *Can. J. Civil Engng,* **10**(3) (1983), 538–48.
14. CAREY, K. L. Ice blockage of water intakes, US Army Cold Regions Research and Engineering Laboratory, Hanover, New Hampshire, report NUREG/ CR-0548 to US Nuclear Regulatory Commission, 1978, 27 pp.
15. CARSTENS, T. Experiments with supercooling and ice formation in flowing water, *Geofysiske Publikasjoner,* **26**(9) (1966), 1–18.
16. COLONELL, J. M. and LIFTON, W. S. Seawater intake design considerations for an arctic environment, *Proc. Cold Regions Engineering Specialty Conf. 1984,* Can. Soc. Civil Engng, Montreal, Quebec, 1984, pp. 275–90.

17. Cox, J. C., Behnke, D. L. and Machemehl, J. L. Design of arctic water-flood intake structures, *Proc. 4th Int. Offshore Mechanics and Arctic Engng Symp.*, Dallas, Texas, Vol. 2, 1985, 135–42.

18. Cox, J. C., Behnke, D. L., Grosskopf, W. G. and Machemehl, J. L. Design and model testing of an arctic seawater intake for sedimentation and wave effects, *Proc. Conf. Arctic 85: Civil Engineering in the Arctic Offshore,* San Francisco, ASCE, 1985, pp. 714–22.

19. Daly, S. F. Trashrack freeze-up by frazil ice (unpublished technical note), US Army Cold Regions Research and Engineering Laboratory, Hanover, NH, 1986, 31 pp.

20. Daly, S. F. Frazil ice dynamics, US Army Cold Regions Research and Engineering Laboratory, Hanover, NH, Monograph 84-1, 1984, 56 pp.

21. Foulds, D. Frazil—the invisible strangler, *J. Amer. Water Works Assoc.,* April 1977, 196–9.

22. Foulds, D. Ice problems at water intakes, *Can. J. Civil Engng,* 1(1) (1974), 137–40.

23. Gerard, R. and Karpuk, E. W. Probability analysis of historical flood data, *J. Hydraulics Div. ASCE,* 105(HY9) (1979), 1153–65.

24. Gevay, B. J. and Erith, H. A. Electric heating of intake trashracks at Twin Falls, Labrador, *Can. J. Civil Engng,* 6(2) (1979), 319–24.

25. Giffen, A. V. Occurrence and prevention of frazil ice blockage of water supply intakes: a literature review and field survey, Ministry of Environment, Canada, publication 1973-W43, 1973.

26. Gosink, J. and Osterkamp, T. E. Measurements and analyses of velocity profiles and frazil ice-crystal rise velocities during periods of frazil-ice formation in rivers, *Annals of Glaciology,* 4 (1983), 79–84.

27. Kuzovlev, G. M. Experience in the design and construction of marine intakes, *Hydrotechnical Construction,* No. 12 (1973), 1151–4 (translated from *Gidrotekhniskoe stroitel'stvo*).

28. Logan, T. H. Prevention of ice clogging of water intakes by application of heat, US Bureau of Reclamation, Denver, Report REC-ERC-74-15, 1974.

29. Matousek, V. Types of ice run and conditions for their formation, *Proc. IAHR Ice Symp. 1984,* Hamburg, Vol. I, 1984, pp. 315–28.

30. Michel, B. Winter regimes of rivers and lakes, US Army Cold Regions Research and Engineering Laboratory, Hanover, NH, Monograph III-Bla, 1971.

31. Morozov, G. A. Methods of counteracting the ice-related problems during the spring at the water intakes at the Southern-Kuzbasskaya SREEP (in Russian), *Trudy koordinatsionnykh soveshchanii po gidrotekhnike,* No. 17 (1965), 95–105.

32. Mussalli, Y. G. Frazil ice control using pneumatic guns, *Proc. IAHR Ice Symp. 1986,* Iowa City, Iowa, Vol. II, 1986, pp. 249–56.

33. Nelson, E. J. Thawing out power, *Instrumentation,* 12(2) (1959), 8–10.

34. Novak, P. and Čabelka, J. *Models in Hydraulic Engineering,* Pitman, London, 1981, 459 pp.

35. Pariset, E. and Hausser, R. Formation and evolution of ice covers on rivers, *Trans. Engng Inst. Canada,* 5(1) (1961), 41–9.

36. Pariset, E., Hausser, R. and Gagnon, A. Formation of ice covers and ice jams in rivers, *J. Hydraulics Div. ASCE,* 92(HY6) (1966), 1–24.

37. PARKINSON, F. E. Frazil ice problems in intakes at Montreal, *Proc. 4th Int. Conf. Cold Regions Engng,* Anchorage, Alaska, ASCE, 1986, pp. 609–18.
38. PERHAM, R. E. Ice sheet retention structures, US Army Cold Regions Research and Engineering Laboratory, Hanover, NH, Special Report 200, 1983.
39. RICHARDSON, W. H. Intake construction for large lakes and rivers, *J. Amer. Water Works Assoc.,* **61**(8) (1969), 365–71.
40. SIGURDSSON, G. The Burfell project: a case study of system design for ice conditions, *Proc. IAHR Symp. Ice and Its Action on Hydraulic Structures,* Reykjavik, 1970, paper 4·0, 18 pp.
41. STAROSOLZSKY, O. Ice and river engineering. In: *Developments in Hydraulic Engineering,* Vol. 3, ed. P. Novak, Elsevier Applied Science, London, 1985, 175–219.
42. STEWART, D. and ASHTON, G. D. Entrainment of ice floes into a submerged outlet, *Proc. IAHR Symp. Ice Problems,* Lulea, Sweden, 1978, Part 2, pp. 291–9.
43. WILLIAMS, G. P. Adhesion of ice to underwater structures, *Proc. 24th Eastern Snow Conf.,* 1967, pp. 82–91.
44. WU, O. C. Offshore seawater intake in the arctic, *Proc. Conf. Arctic 85: Civil Engineering in the Arctic Offshore,* San Francisco, ASCE, 1983, pp. 551–6.

Chapter 3

THE INTERFACE BETWEEN ESTUARIES AND SEAS

D. M. McDowell

*Emeritus Professor of Civil Engineering,
University of Manchester, UK**

NOTATION

Symbols used only once are defined locally in the text. To achieve homogeneity, some symbols have been changed from those used in source publications.

a	wave amplitude
B	$c \cdot c_g$
b	separation between wave rays
C	Chezy coefficient of friction
C_r	Courant number relative to water; ≤ 1 or 1 for numerical stability of explicit schemes
c	wave celerity
c_g	wave group celerity; c_{ga} is absolute, c_{gr} is relative to water
D_m	energy dissipation rate at bed
D_x, D_y	depth-averaged coefficients of dispersion
d	water depth ($= h + \eta$)
E	wave energy per unit plan area
e	wave energy per unit width $= (1/2)\rho\mathbf{u}_w^2 + \rho g z + p$

* Present address: 13 Powis Villas, Brighton, East Sussex BN1 3HD, UK.

F	$F(z) = \cosh k(z + d)/(\cosh kd)$
G	see local definition
g	gravitational force per unit mass
H	wave height ($2a$ for simple harmonic waves)
h	level of bed below reference plane
i	$(-1)^{0.5}$
K	wave number $= 2\pi/L$; $K^2 = k^2(1 + \delta)$; $\omega_r^2 = gK\tanh(Kd)$
K_j	component of K in j- (or y-) coordinate direction
k	wave number $= 2\pi/L$ from linear wave theory; $\omega^2 = gk\tanh(kd)$ for small-amplitude waves
L	wavelength
M	see local definition
m	mass of sediment per unit area available for transport
n	coordinate normal to direction of wave propagation
p	wave energy propagation velocity $= \langle \int_{-h}^{\eta} \mathbf{u}_w^2 e\,dz\rangle/\langle\int_{-h}^{\eta} e\,dz\rangle$
p	pressure (Section 6); dimensionless friction factor (Section 9)
p, q	components of discharge per unit width in x- and y-directions
R	intrinsic or kinematic wave group velocity
S	(Section 6) source or sink term for water or solids in motion
S	(Sections 8, 9) see local definitions
S_{ij}	radiation stresses
s	coordinate in direction of wave propagation (Section 8)
s	ρ_s/ρ, ratio of sediment density to water density
T	wave period
T_s	mass rate of transport of sediment per unit width
t	time
U, V	components of depth mean flow speed in x- and y-directions
u, v	components of flow speed in x- and y-directions at elevation z
U_*	shear velocity $(\tau_0/\rho)^{0.5}$
w	component of water velocity in z-direction
x, y	horizontal coordinates
z	vertical coordinate, positive upwards
α	see local definitions
β	surf parameter
γ	see local definition
δ	$(\partial^2 a/\partial x_i^2)/a$
Δ	$\partial/\partial x$, $\partial/\partial y$, $\partial/\partial z$
Δ^2	$\partial/\partial x^2$, $\partial/\partial y^2$, $\partial/\partial z^2$
ε	eddy mixing coefficient (eddy viscosity)
ζ	sediment concentration at elevation z

η	elevation of water surface above reference plane. N.B. In combined flows, η includes tidal rise and fall and set-up or set-down due to waves as well as any instantaneous variation during passage of a wave
θ	angle between orbital wave motion and normal to current direction
κ	von Karman's constant (~ 0.4)
μ	u_{cw}/u_c
v	kinematic viscosity
ρ	water density
ρ_s	density of solids on the bed
τ_0	shear stress at the bed
τ_{bx}, τ_{by}	components of shear stress at the bed
τ_{sx}, τ_{sy}	components of shear stress at the surface
ϕ	$\phi(x, y, z, t)$, velocity potential
ψ	$\psi(x, y, t) \cdot f(z)$, velocity potential in depth-averaged flow
ω	angular frequency of waves $= 2\pi/T$
ω_a	absolute angular frequency with reference to bed
ω_r	angular frequency relative to moving water mass
χ	see local definition
∇	$\partial/\partial x, \partial/\partial y$
∇^2	$\partial/\partial x^2, \partial/\partial y^2$

Subscripts

b	indicates condition at breaking of wave
c	component due to current
cr	critical value, e.g. for start of sediment transport
cw	component due to combined waves and currents
i, j	indicate x- and y-directions in tensor notation
m	represents c, w and cw
n	normal to wave propagation or current direction
p	parallel to current direction
w	component due to waves
Overbar	indicates depth-averaged value
$\langle \ \rangle$	indicates average over a wavelength

1 INTRODUCTION

Flow of water and sediments in rivers is one-way, under the action of gravity. When rivers approach the sea they come under the influence of

tidal action which initially causes periodic variations in the rate of flow
and, further down the seaward course, periodic reversals imposed on the
flow. Somewhere within the zone of tidal influence the fresh river waters
encounter saline water and from there onwards they are influenced by
differences in density over the depth and in the seaward direction. These
density gradients superimpose a differential motion on the net seaward
movement, seawards at the surface and landwards at the bed. The overall
effect depends on the relative magnitudes of tidal influx and river discharge during a tidal cycle over unit width of the estuary and on the degree
of turbulent mixing.[42]

Sediment movement is seawards in the river but the differential water
movement caused by density gradients can give rise to landward
movement of sediment in the tidal estuary. This is likely to vary seasonally;
in many rivers there are times when fresh water flow is large enough to
carry sediment out to sea. At other times, sediment accumulates within the
estuary or at its mouth so that there is progressive land-building and
seaward extension of the neighbouring coast. Near the sea-face, wave
action becomes important in inducing additional currents and in initiating
or enhancing bed movement by oscillatory motion.

The relative importance of the forces of tidal currents and waves may
vary according to the season or local weather. Where tidal energy is
relatively high, estuaries tend to widen exponentially towards their mouths
so that the effect of river flow weakens. At times of low river flow the effect
of fresh water discharge may be so small that sediment transport is not
significantly affected by it; tidal currents dominate. Where tidal energy is
low, the combined effect of river discharge and tide-induced currents may
be insufficient to transport sediment. This is often the case in the region
outside the line of the coast. Wave action is then necessary for sediment
movement to occur at all. Without it, sediment would accumulate and the
coastline would migrate seawards. With wave action, sediment movement
can take place under the influence of quite weak tidal or density currents.

Wave action at the sea-face of an estuary is likely to be affected by
shallow water which can cause refraction, leading to local variations in
wave height. These can become large enough to induce an appreciable
spread of wave energy by diffraction. In water shallow enough to induce
breaking, gradients in the flux of momentum give rise to radiation stresses
which result in set-up and set-down of local mean water level and also
cause wave-induced currents. In the case of estuaries which have high tidal
energy, strong current gradients in plan may cause additional refraction of
waves. Variations of wave height normal to the direction of propagation
can result from any of these causes. They give rise to diffraction of waves.

The whole situation is one of interacting waves and currents with a high degree of non-linearity due to energy dissipation by friction and wave breaking as well as to convection. The complexity of conditions at the sea-face of estuaries is such that full analysis is not possible. However, there have been advances in physical understanding and in numerical

TABLE 1
DETAILS OF THREE ESTUARIES

	Santos (Brazil)	Rio Pungue (Mozambique)	Mersey (England)
Tides			
$\dfrac{K_1 + O_1{}^a}{M_2 + S_2}$	0·36	0·023	0·056
Mean spring rise/fall	1·1 m	5·7 m	8·1 m
Distance inland to limit	20 km	> 100 km	24 km
Typical storm surge	1·3 m	> 1 m	1·1 m
Tidal prism at mouth	$1·9 \times 10^7 \text{ m}^3$	$7·0 \times 10^8 \text{ m}^3$	$8·8 \times 10^7 \text{ m}^3$
Offshore slopes			
Distance to −20 m contour	11 km	33 km	24 km
−50 m contour	54 km	130 km	60 km
River discharges			
Typical flows	300 m³/s max.	900 m³/s max.	1 200 m³/s max.
	90 m³/s min.	80 m³/s min.	56 m³/s mean
Dominant winds			
	1978/80	1946–1955	1969/70
Directions	60–120°	68–158°	270–315°
Speeds	Mean 3·2 m/s	4·0–5·4 m/s	> 11 m/s
% time	48%	60%	12% of year
Directions	180–210°	Occasional	270–315°
Speeds	Mean 3·3 m/s	tropical	> 22 m/s
% time	18%	cyclones	2·5% of year
Normal maxima	22 m/s from NE	8–11 m/s (Jan.)	~ 30 m/s
Extreme	> 28 m/s (gust)	> 23 m/s	> 32 m/s
Waves			
			12% 2·5%
Significant height	1·7 m	~ 2 m	> 2·5 m > 5·4 m
Maximum height	4·6 m recorded (bay entrance)	7 m recorded (pilot vessel) 4·6 m max. (Macuti channel)	> 4 m > 8.7 m
Period	$T_{50} \sim 10·5$ s	$T_s \sim 7$ s. (a few at $T > 19$ s)	$T_s \sim 7$ s 9 s
Directions	120–200° (open sea) 155–165° (Santos bay)	50–160°	270–315° (whole year)

^{a}K$_1$, O$_1$, M$_2$ and S$_2$ are tidal constituents. See reference 42, Ch. 1.

modelling over the past decade so that modelling can be of considerable help in solving real problems. These advances are the subject of thi chapter.

2 SOME TYPICAL CASES

The great variability of estuary/coastal interfaces can best be demon strated by means of examples. Three have been chosen to illustrate a range of conditions that may be encountered. The main features of these estu aries are summarised in Table 1.

The Santos estuary, Brazil (Fig. 1), has weak tides and a small fresl water outflow. Wave action is seldom severe but there is a swell witl period of order 11 s for much of the time. This causes oscillator movement of sediment on the bed of the bay but cannot give rise t appreciable net movement. When combined with the tidal currents, whicl are too weak to move much sediment by themselves, there is enougl sediment movement to cause serious problems in the approach channel t the port of Santos. The net sediment movement is affected greatly b salinity gradients which cause near-bed circulation to differ from that o the near-surface layers. Refraction causes waves to enter the bay througl a narrow sector between 155° and 165°. The wave energy is thus greates in the western part of the bay, causing a bar to form at the entrance to th Sao Vicente estuary. The bed in that part of the bay consists of fine sand The entrance to the Santos estuary in the north-east part of the bay i sheltered from wave action and the bed of the eastern part of the bay consists of sandy silt.

The Pungue estuary, Mozambique (Fig. 2), has moderate tides and fresl water flow. Sediment is moved easily by tidal currents. There are extensive shallows outside the mouth of the estuary and the channels between then are in a state of dynamic equilibrium. Wave action has considerable effec on nearshore sediment movement. The moderately long waves break on the outer shoals at low tide and have a major effect on their location and extent. Attenuated waves reach the shore, particularly near the times o high tide, and cause considerable alongshore transport towards the mouth of the Pungue estuary from the east. Sediment carried to the outer bars by the strong ebb currents may be driven to the shore by wave action particularly during the rising tide, and then carried back to the mouth of the estuary by the alongshore wave-induced currents.

The Mersey estuary, England (Fig. 3), has strong tides and wave action but relatively small fresh water flow for much of the time. The quantity

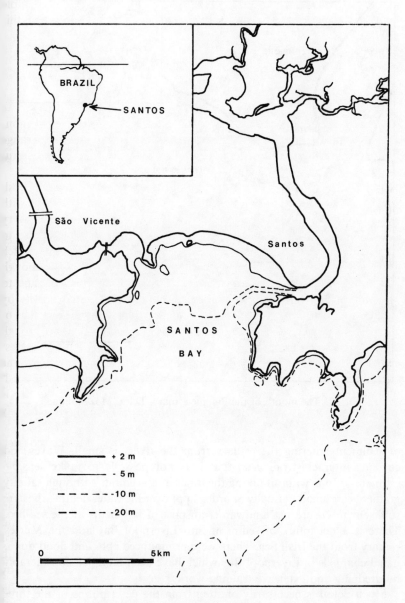

FIG. 1. Santos Bay, Brazil

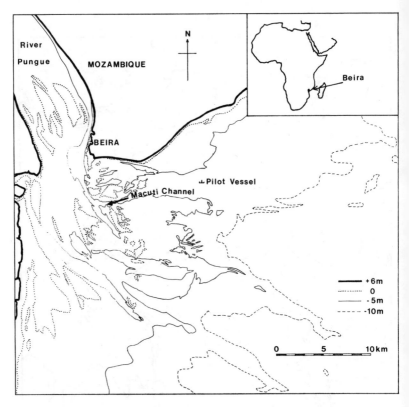

FIG. 2. The mouth of the Pungue estuary, Beira, Mozambique.

of sediment entering the estuary from the rivers is small. Waves and currents interact strongly so that it is not possible to make sensible estimates of one without taking the other into account. Although salinity gradients are small—usually of order 1 ppt over the water depth—they are sufficient to ensure net landward transport of sediments at the sea-bed. There is a net influx of sediment into Liverpool Bay and the Mersey estuary from the Irish Sea. There were pronounced ebb- and flood-dominated channels in Liverpool Bay which have been modified considerably by dredging and extensive training works.

Much detail is necessarily omitted from the descriptions of the three cases, but there is enough to show the common features which have to be analysed before engineering problems can be solved. They are:

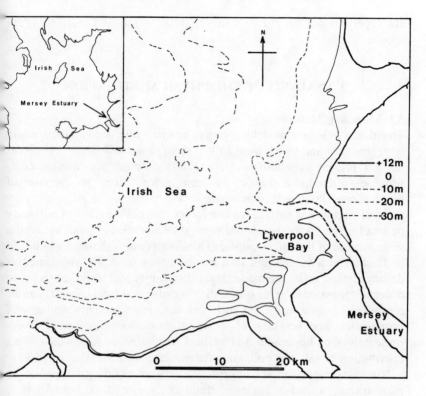

FIG. 3. Liverpool Bay and the mouth of the Mersey estuary, England.

(1) The directions of net sediment transport differ fundamentally from those of depth-averaged movement of water by tidal flow. Outside the zone of breaking waves, this is mainly due to salinity gradients.

(2) Sediment transport by tidal currents is affected by the presence of wave action. Its rate may be increased but bed movement also occurs in places where currents alone would be too weak to initiate it.

(3) Waves are refracted as they approach the coast by shallow-water effects and by passage through moving water. Where tidal currents are strong the effect on the pattern of waves in plan can be considerable. In such cases the interaction of tidal currents and waves cannot be ignored.

(3) Refraction and diffraction of waves give rise to local variations of wave energy and momentum flux which, in shallow water, cause wave-induced circulation of water and sediment. These wave-induced currents

are superimposed on currents due to tidal action and meteorological effects.

3 ANALYSIS OF THE PHYSICAL SITUATION

3.1 Modelling Methods

The ideal would be a modelling system in which all these variables would be reproduced and which would allow them to interact freely. That is far beyond present capabilities. All methods of modelling involve compromise—often quite drastic—between major forces. In the case of physical modelling, reduction in size results in increase in relative importance of viscous and surface tension forces whereas the effect of both may be small in the prototype. Estuaries must be modelled at small horizontal scale because of their physical size. These scales are such that flows would be laminar in a model built to a homogeneous scale. To overcome this defect, vertical scales are made larger than horizontal scales, often by an order of magnitude.[45] This is essential in order that tidal currents can be reproduced satisfactorily; but such scale distortion prevents similarity of propagation and breaking of waves. It can also cause severe misrepresentation of horizontal and vertical mixing processes[45] which affect distribution of salt and pollutants (particulate, solute or thermal).

Numerical models represent prototype conditions on a grid of points in plan so that, instead of having continuously varying solid boundaries as in nature and in physical models, boundary values are only stated at intervals. Similarly, variations of parameters with time are estimated at intervals that are short compared with a tidal period but in some cases are long compared with wind-induced waves and swell and are very long compared with turbulent fluctuations. Modelling of wind-induced waves and of tides requires different compromises to be made. Tides are 'long' waves in 'shallow' water in which vertical accelerations are very small compared with gravitational force per unit mass. In the case of wind-generated waves and swell, vertical accelerations cannot be ignored. Tide- and wind-generated waves are each modified by passage through 'shallow' water. In the absence of significant tidal currents, wave energy E per unit plan area is propagated at the wave group celerity c_g.

Group celerity changes as waves pass into shallow water or encounter moving water, causing wave height to change. Local changes in wave speed cause directional changes in the wave front (refraction). Changes in wave height along a wave front cause wave energy to spread laterally (diffraction). When waves move onto a sloping shallow bed and approach the point of breaking, they slow down and exert a force on the bed with a resulting change in flux of momentum. Radiation stresses then cause changes in mean water level and generate wave-induced currents.

A further complication is that waves and tides consist of an infinite range of components. In the case of tides, the components of the tide-generating forces can be measured and their effect predicted with precision. However, shallow water and reflection from coastlines cause local changes which are usually orders of magnitude greater than the effect of direct tide-generating forces. These local effects can be predicted with quite high accuracy. Seasonal effects, which can cause changes in mean sea level with periods of the order of a year, can be measured over several years and mean variations found. Meteorological effects also cause short-term variations with periods of the order of days. Seasonal and meteorological effects must necessarily have large random components which prevent accurate forecasting. Tides can be predicted fairly well in regions where the astronomical tide causes large shallow-water tides. Where the shallow-water tide is small, meteorological and seasonal effects can dominate the rise and fall of water, and prediction of tidal action lacks any precision.

In the case of wind-generated waves, the wave climate at any locality can be represented as a series of spectra, each of which describes the distribution of wave energy with frequency and direction. Coastal topography and offshore contours limit the directions at which waves can approach a coast. Long waves are refracted in deeper water than short waves, so that the climate at any coastal locality will, in general, differ from that in deep water. Large waves also break in deeper water than smaller waves of the same period. Offshore shoals that are typical of many estuary mouths can thus cause filtering of some components of the incident wave spectrum (not necessarily those exceeding a particular height or a particular wavelength or period). At a simplistic level, those components that would break might be assumed to be partially filtered out, but that assumes linear separation of wave components, which is not applicable under conditions of wave-breaking. Finally, at the mouths of estuaries tidal currents are usually strong. Waves propagating through moving waters are refracted in the process to such an extent that the inshore wave climate in the presence of currents is different in major respects from that which would occur in their absence.[33]

The choice of method of analysis of estuarial behaviour depends on the purposes of study. Morphological changes resulting from sediment transport are central to understanding of physical behaviour and of reaction to imposed changes due to dredging, dumping of solids, reclamation or training works. Another frequent requirement is for knowledge of dispersion of pollutants due to discharge from outfalls or dumping near coasts.

Although much progress has been made in understanding the mechanics of sediment transport, there is still no satisfactory formulation that enables forecasts of transport rates to be made within an accuracy of say 50%, even when the determining parameters such as grain and flow properties are closely specified. It is often argued that, because of this deficiency, there is no point in spending a great deal of effort in increasing the accuracy of knowledge of flow field and wave climate. This can be seen in perspective when it is realised that the sediment transport rate depends on a high power of the depth-mean flow speed—typically its 4th or 5th power.

Errors in estimating a depth-averaged flow speed U of 1 m/s would then have the following effects (given as percentage errors in estimates):

U	-20	-10	$+10$	$+20$
U^4	-59	-34	$+46$	$+107$
U^5	-67	-40	$+61$	$+149$
$(U^5)\% - (U^4)\%$	-8.2	-6.5	$+15$	$+42$

Thus a $+10\%$ error in determining the flow speed would result in a 46% error in sediment transport rate if it is dependent on U^4 or a 61% error if dependent on U^5, whereas the difference between these values is only 15%. It is evident that, in this case, quite small changes in flow field can have a significant effect on sediment transport rates and that errors in formulation of sediment transport rates may be no more serious than errors in determining velocity and shear stress vectors. Because the problem is so highly non-linear it is not possible to generalise about the relative importance of accuracy in determining hydrodynamic behaviour and in formulating the resultant rate of sediment transport.

3.2 Calibration and Verification of Models

The complexity of conditions at the mouths of estuaries is so great that analysis of the whole situation, with interactions between the various components, is not possible. The best that can be done at present is to identify the major components of a given situation, to separate them as far as necessary to reduce analysis to a manageable level, and to ignore non-linear interaction wherever possible. Under these conditions, carefully chosen field measurements are essential as a check on potential accuracy.

The results of numerical analysis nowadays can be detailed and picturesque. It is sometimes claimed that they can give more, and more accurate, information than could be obtained in the field, and that many field measurements are so imprecise that they should be treated with suspicion until they have been verified by computation. Acceptable field measurements are those of water level, subject to seasonal and meteorological variations which may affect comparison with computations based on standardised conditions. Less accurate are soundings, which must be sufficiently detailed to account for variations due to bed forms (sand waves, dunes and ripples) and must be completed during a time span short compared with changes in bed level, including local bed forms. This is a tall order in a highly mobile estuary. Measurement of water velocities is slow, expensive and prone to error when the object is the usual one of establishing vertical profiles of flux of water and/or sediment. The greatest difficulty is to establish the exact position of measurements in the water mass relative to the bed. Measurement of sediment transport rates is the most difficult of all. Movement in suspension is relatively easy; the water can be sampled. However, reasonably accurate rates of sediment transport can only be measured by sampling close to the bed. Distance from the boundary must be known quite accurately, but the boundary may be wavy, with amplitude of the order of a metre, and mobile. The rate of bed-load transport can never be ignored except in cases where the bed consists only of fine silts and clays. On the other hand, it cannot be measured directly without interference with the local flow. The whole problem is greatly complicated by the presence of waves. Indirect measurements using natural or artificial tracers have proved to be effective in some cases.[25] Field measurements can only be representative of the conditions under which they were obtained, including random seasonal and meteorological effects. It is seldom easy to identify and model these conditions.

Field measurement methods have advanced considerably. Radar methods can be used to record wave directions and surface velocity vectors at suitable locations.[53,69] They can provide instantaneous measurements over a large area which only need to be supplemented by very few point measurements of wave spectra and velocity/depth profiles. They can then give a comprehensive picture for comparison with computed values. Ultrasonic methods capable of obtaining instantaneous vertical profiles of velocity are being developed. In certain cases they may also give information about concentrations of sediment in suspension.[18,66] It is thus becoming more practical to verify the performance of numerical models by means of field measurements.

The most rapid developments are occurring in numerical methods and the rest of this text is therefore devoted to them.

4 INTRODUCTION TO NUMERICAL ANALYSIS OF INTERACTIONS BETWEEN ESTUARIES AND SEAS

Numerical models represent conditions at finite intervals of space and time. For this reason alone, they can never simulate the full Navier–Stokes equations, though they can be used to solve their approximate space- and time-averaged forms. Estuarial processes have an extremely wide range of time scales; from fractions of a second to a few tens of seconds for turbulent diffusion; from under 4 s to over 20 s for wind-generated waves and swell; from a few minutes to an hour or two for interfacial waves in density-stratified flows; from 12 h to 25 h for primary tides but with shorter-period harmonics in shallow water; from several days to some months for morphological changes. Major solar–lunar tide-generating forces vary over an approximately cyclic time of about 19 years.

To be successful, a numerical model must be based on a time interval that is short compared with the shortest time event that has to be fully modelled. In the course of analysis, harmonics of initial and boundary waves are likely to be generated. The time interval chosen will determine the cut-off frequency for computation of such harmonics. Many shorter-period effects can be represented with sufficient accuracy by space- or time-averages. Thus any attempt to model the complete equations for morphological changes would require a time-scale based on tides and/or waves which would be wildly uneconomic.

There are several ways of overcoming this dilemma. One is to ignore the detailed analysis of an event but to estimate its overall effect; depth- or cross-sectional averaged conditions can replace local values; friction coefficients can be used in place of determination of local shear stresses; representative monochromatic waves can be used in place of wave spectra; propagation of a property such as wave energy can be used instead of detailed analysis of wave profiles.

Another approach is to break the problem down into parts which can be analysed separately. This ignores non-linear interactions between them, but can be effective when these are weak. Tidal currents, offshore waves, nearshore waves and local conditions of waves and currents can be studied in sequence and the results used to estimate local rates of sediment transport or dispersion.

Yet another approach is to consider that temporal changes are made up from the sum of several effects which can each be described by appropriate forms of equations, such as elliptical, parabolic or hyperbolic. Methods of solution suitable for each form can be used and the results combined during each computational time interval.

Until quite recently, the only possible procedures were to separate currents caused by tides and river flows from effects due to waves, including wave-induced currents. A suite of numerical models would be used according to the problem being studied. Recent developments allow some

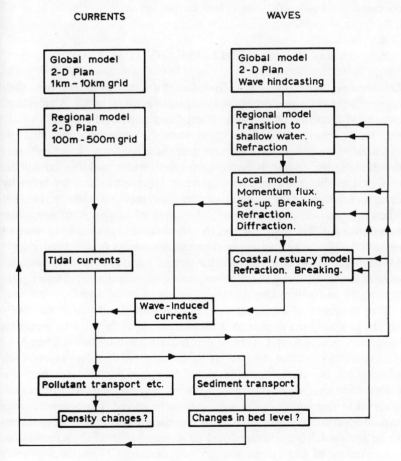

FIG. 4. A modular scheme for modelling estuarine interfaces.

aspects of wave/current interactions to be modelled. This is a significant advance which will be followed by further developments as costs of computing fall.

A scheme for modelling the entrance to a well mixed estuary on a modular basis is shown in Fig. 4. In modelling a partially mixed or a stratified system, a width-averaged two-dimensional model can be used within the estuary but this is not possible in a wide system; in such a case there is no alternative to a three-dimensional or a layered two-dimensional model.[23,24,37]

Global models which reproduce tidal currents or wave climate are well developed and will not be described further herein.[31,55]

5 REGIONAL MODELLING OF TIDAL FLOWS

Estuaries discharging through alluvial coasts widen considerably as they approach the sea in response to the increase in net tidal influx. An idealised estuary with uniform depth and velocity amplitude over its length would necessarily do so exponentially,[51] and many real estuaries approximate to this ideal. As a result, there is room near the coast for separate ebb- and flood-dominated channels to develop—the former conditioned by the direction of the major estuary channel at the mouth, and the latter by coastal currents and the momentum of water remaining in the ebb channel when the tide is low at the mouth. Modelling of estuary mouths must be two-dimensional in plan; some early models included an area of sea and estuary mouth linked to a one-dimensional model of the tidal river.[52] Modelling of dispersion is an essential part of a study of seas at the mouths of estuaries. This topic has been reviewed recently by Holly[27] and Sauvaget[58] and will not be discussed further in this chapter.

Two-dimensional depth-averaged models were developed in the 1960s and have since been refined to such an extent that they can be used as a matter of routine in the study of depth-averaged tidal flows. They have been extended to three dimensions by dividing the flow into layers, each of which is two-dimensional in plan.[38] Separate models have been developed for analysis of different problems, and this separation has been dictated by the nature of the equations to be solved. Major laboratories can offer various combinations of such models according to the problems to be analysed. The equations used have been based on the equations of conservation: of mass, of momentum flux, of energy. Methods of solution have been based on explicit and implicit finite difference models,[1] on finite elements,[13] on mixed finite difference/finite element approaches,[37] and on the method of characteristics.[36] Each has advantages and drawbacks, and each has been refined considerably over the past decade.

The starting point of all these methods is the 'shallow water' equations based on the assumption that pressure is hydrostatic and that the water is well mixed vertically. They describe conservation of momentum, of mass, and of constituents:

Conservation of momentum:

$$\frac{\partial U}{\partial t} + \frac{U\partial U}{\partial x} + \frac{V\partial U}{\partial y} - fV + g\frac{\partial \eta}{\partial x} - M\left(\frac{\partial^2 U}{\partial x^2} + \frac{\partial^2 U}{\partial y^2}\right) + g\frac{d\partial\rho}{2\rho\partial x}$$

$$+ g\frac{U(U^2 + V^2)^{0.5}}{C^2 d} - \frac{\tau_{sx}}{\rho d} = 0 \tag{1}$$

$$\frac{\partial V}{\partial t} + \frac{U\partial V}{\partial x} + \frac{V\partial V}{\partial y} - fU + g\frac{\partial \eta}{\partial y} - M\left(\frac{\partial^2 V}{\partial x^2} + \frac{\partial^2 V}{\partial y^2}\right) + g\frac{d\partial\rho}{2\rho\partial y}$$

$$+ g\frac{V(U^2 + V^2)^{0.5}}{C^2 d} - \frac{\tau_{sy}}{\rho d} = 0 \tag{2}$$

Conservation of mass of water:

$$\frac{\partial \eta}{\partial t} + \frac{\partial(dU)}{\partial x} + \frac{\partial(dV)}{\partial y} - S = 0 \tag{3}$$

Conservation of constituents (solutes, solids), for each constituent:

$$\frac{\partial(d\bar{P})}{\partial t} + \frac{\partial(dU\bar{P})}{\partial x} + \frac{\partial(dV\bar{P})}{\partial y} - \frac{\partial(dD_x(\partial\bar{P}/\partial x))}{\partial x} - \frac{\partial(dD_y(\partial\bar{P}/\partial y))}{\partial y}$$

$$+ [K]d\bar{P} - \bar{S} = 0 \tag{4}$$

U, V are components of depth-averaged velocity in the x,y directions; f is the Coriolis coefficient, $2\psi\sin\phi$; ψ is the angular velocity of Earth; ϕ is the angle of latitude; η is elevation of the water surface above reference plane (mean water level); h is level of the solid boundary below reference plane; $d = (h + \eta)$; ρ is water density; M is coefficient of horizontal diffusion of water; C is Chezy friction coefficient; τ_{sx}, τ_{sy} are components of shear stress at the surface due to wind and waves; τ_{bx}, τ_{by} are components of shear stress at the bed; g is gravitational force per unit mass; S is a source or sink term; $\bar{P}$ is a depth-averaged concentration for each constituent present (by volume); D_x, D_y are coefficients of depth-averaged horizontal dispersion for each constituent; $[K]$ represents a matrix describing reactions between constituents; $\bar{S}$ is a depth-averaged source or sink term for each constituent.

If any of the constituents is in such high concentration that it causes appreciable differences in density, a further equation of state is required, e.g.

$$\text{For salts:} \qquad\qquad\qquad \rho = \rho(s) \qquad\qquad (5)$$

$$\text{For solids in suspension:} \; \rho = \rho(c) \qquad\qquad (6)$$

These equations are based on the following assumptions:

1. Flow is well mixed
2. Fluid is incompressible
3. Vertical accelerations can be neglected when compared with gravity
4. A quadratic friction law is valid
5. Terms are averaged over a time interval that is long compared with turbulent fluctuations but short compared with tidal period

Because of the assumptions on which these equations are based, they cannot take proper account of buoyancy effects such as would occur in partially-mixed or stratified estuaries and they cannot reproduce the consequences of rapid variations in bottom topography.

Early numerical models used explicit or implicit methods of solution at each time step.[1] Explicit methods are limited to time steps determined by the Courant number:

$$C_r = \Delta t [gd(1/\Delta x^2 + 1/\Delta y^2)]^{0.5} < 1 \qquad\qquad (7)$$

in order to ensure numerical stability. The explicit method is simple to set up. Due to this and its acceptable accuracy it has been widely used despite the need to adopt short time steps to keep within the Courant criterion. Implicit methods are inherently stable and are usually operated at Courant numbers higher than unity. Usual practice has been to divide the time interval into two halves and to solve the equations for the x- and y-coordinate directions in one half each. This simple Alternating Direction Implicit method (ADI) is based on linear interpolation across the time step. It has been found that this can give good results at low Courant numbers (< 16) but that errors become rapidly more serious as the Courant number is increased to > 20. There is the added danger that there is no indication (through instability) that the results are wrong. It has been shown that the ADI method in its simplest form is equivalent to iterative solution of elliptic equations—but using only one iteration.[6] To obtain reasonable answers at high Courant numbers, many iterations would be required and the economy of the method would be lost.

An alternative approach to solution of the equations has been developed which overcomes many of the problems of solving the equations using high Courant numbers (Benque *et al.*[6,7]). Equations (1)–(3) are rearranged:

$$
\begin{vmatrix} \dfrac{\partial p}{\partial t} \\[2mm] \dfrac{\partial q}{\partial t} \end{vmatrix} + \begin{vmatrix} \dfrac{\partial (pU)}{\partial x} + \dfrac{\partial (pV)}{\partial y} \\[2mm] \dfrac{\partial (qU)}{\partial x} + \dfrac{\partial (qV)}{\partial y} \end{vmatrix} + \begin{vmatrix} gd\dfrac{\partial \eta}{\partial x} \\[2mm] gd\dfrac{\partial \eta}{\partial y} \\[2mm] \dfrac{\partial d}{\partial t} + \dfrac{\partial p}{\partial x} + \dfrac{\partial q}{\partial y} \end{vmatrix} - \begin{vmatrix} fq + \tau_{sx}/\rho \\[2mm] fp + \tau_{sy}/\rho \end{vmatrix} + \begin{vmatrix} \dfrac{\tau_{bx}}{\rho} \\[2mm] \dfrac{\tau_{by}}{\rho} \end{vmatrix}
$$

$$
\begin{array}{ccccc} \text{I} & \text{II} & \text{III} & \text{IV} & \text{V} \end{array}
$$

$$
-\begin{vmatrix} \left(\dfrac{\partial}{\partial x}\left(M\dfrac{\partial p}{\partial x}\right) + \dfrac{\partial}{\partial y}\left(M\dfrac{\partial p}{\partial y}\right)\right) \\[4mm] \left(\dfrac{\partial}{\partial x}\left(M\dfrac{\partial q}{\partial x}\right) + \dfrac{\partial}{\partial y}\left(M\dfrac{\partial q}{\partial y}\right)\right) \\[4mm] {} \end{vmatrix} \begin{matrix} = 0 \\[4mm] = 0 \\[4mm] = 0 \end{matrix} \qquad \begin{matrix} (8) \\[4mm] (9) \\[4mm] (10) \end{matrix}
$$

$$
\text{VI}
$$

where p and q are the components of discharge per unit width in the x- and y-directions respectively. The authors pointed out the physical meanings of the groups:

I. Local flow acceleration: local variation of momentum with time
II. Spatial acceleration
III. Conservation of mass and momentum change due to wave propagation
IV. Momentum sources and sinks due to Coriolis force and wind stress
V. Momentum sink due to friction at the bed
VI. Horizontal diffusion of momentum

The equations represent the superposition of three different kinds of operation:

(1) Advection (groups I and II)
(2) Diffusion (groups I, IV and VI)
(3) Propagation (groups I, III and V)

The first of these is kinematic, the second is a process of spreading, whil
the third is dynamic. The authors described a method of solving th
equations using a method appropriate to each operation.

In the finite-difference scheme, each time interval was divided into thre
parts. One operation was carried out during each part. The methods o
computation used were:

— for the advection step the method of characteristics, each coordinat
  direction being treated separately;
— for the diffusion step an alternating direction implicit method, wit
  the output from the advection step as initial values;
— for the propagation step an alternating direction iterative method
  which required a large number of iterations at high Couran
  numbers; typical values were 15 to 50 iterations at Courant number
  of 20 to 40.

It is noteworthy that calculation of depth h only occurs in the propaga
tion step. This leads to simplification and considerable economy in com
puting the other steps. The terms making up the propagation step wer
rearranged in finite difference form and linearised to give a two-dimen
sional equation for d in terms of previously calculated values of U and V
Because the resulting equation was symmetrical it was possible to solve i
using an alternating-direction operator at high Courant numbers rathe
than solving the full two-dimensional equation. Despite the large numbe
of iterations, reasonable economy was achieved by virtue of the higl
Courant number used and the relative simplicity of the equations. Th
method was adapted so that allowance could be made for treatment o
uncovering of tidal flats.

When applied to real problems, Courant numbers as high as 95 wer
used. One of several examples described by Benque et al.[6,7] was a study o
currents at the mouth of the La Canche estuary. Input parameters at th
model boundary were obtained from a model of the English Channe
based on a square grid with 10 km sides. The grid in the La Canche mode
had 100 m sides. The flow field included major eddying during the tida
cycle. Currents measured throughout a tidal cycle at four locations acros:
the mouth of the estuary showed good similarity with computed speed:
and directions.

6 MODELLING OF TIDAL CURRENTS AT ESTUARY MOUTHS

The shallow-water, nearshore regions of estuary mouths are usually
characterised by extensive deposits of sand and/or silt which are re-worke(

by tidal currents to form channels, bars and shoals. Large areas which are immersed at high tide may become exposed as water levels fall. Towards low tide the ebb current is confined largely to the main ebb channels, but as tide levels rise water enters the estuary on a widening front. Flows in outer estuaries tend to be well mixed due to the strength of tidal currents, though large river flows or weak tides may cause a degree of stratification.

Methods of analysis must cater for the transition between regions with comparatively mild bed slopes and moderate depths and regions with steep slopes, a wide range of depths and drying banks. Waves have a big influence on formation of shoals and sediment transport. To achieve practical economies it is necessary, though incorrect, to decouple the calculation of waves and wave-induced currents from that of tidal currents.

Models for study of tidal currents in nearshore regions need to resolve finer detail than is necessary in regional models, and they have to take drying banks into account. One way of dealing with fine detail is to use a relatively coarse grid for the whole region and to use a fine grid nested within it for study of nearshore detail. This can work well if shorelines and contours are not very convoluted and if there are not large drying areas. The usual procedure has been for the coarse-grid model to generate boundary conditions for the fine-grid nested model, as was done for the La Canche estuary described above.[6]

Modelling of drying regions poses computational and numerical problems. Zero depth is anathema to computers and they have to be instructed to recognise and avoid it. This necessitates checking all marginal grid points whenever depths approach zero. The problem is compounded by the need to decide when a grid cell should be regarded as dry. It may take several time steps for a cell in a coarse grid to change from all wet to all dry, and during that time neighbouring cells will be undergoing similar changes. Numerical errors through interpolation are inevitable and can lead to failure to satisfy conservation requirements. The problem is much less severe if the computational grid can be made as fine as practical, and it can be eased further if the grid (non-rectangular) has sides nearly parallel to the drying line.

Weare[67] has shown that the ADI method can lead to errors in calculation of velocity profiles, particularly when boundaries are irregular. In a recent paper, Stelling et al.,[62] examined this further and showed that, in many cases, irregular boundaries limit the area that can be covered during each sweep in the x- or y-direction. This is likely to be a major factor in regions where there are extensive drying areas. In such cases, use of a high Courant number—greater than about 5·7—will result in incorrect speed of propagation of the computed tide. They also showed that the method of

checking each grid point for drying is important. Three methods wer
compared: one in which the local water depth is checked at every gric
point where velocity is calculated, and two in which the local water dept
is checked at every point where water surface level was calculated. The
showed that these methods produced different propagation speeds an
different drying areas, particularly during falling tides. The first metho
appears to be the most accurate; it is used when only water levels an
velocities are calculated. One of the other methods is needed when trans
port of dissolved material is being studied. The authors concluded that, fc
general modelling, ADI or fully implicit methods are preferable to explic
methods despite these inaccuracies, partly because of the freedom fror
depth-dependent stability criteria. The Courant number for explicit con
putations must be based on the greatest depth occurring in the systen
whereas that quoted for the ADI method can be based on the mean deptl
   One alternative method works particularly well with explicit method
It makes use of Distributed Array Processors (DAP)[43] in association wit
a host computer. The DAP consists of a set of 64×64 array processor
which can be programmed to execute the same calculation sequenc
simultaneously in each processor. It can be used in several ways; fo
example, an area can be divided into several blocks each covered by
62×62 mesh. The DAP can join the blocks dynamically during a run s
that calculations are performed as if the whole region was covered by
continuous mesh. Another use is in patching models of local areas base
on fine grids within regional models based on coarser grids. Several sucl
patches can be run simultaneously, each representing a different localit
within the region. DAP can also be used to build up 2-D layered and 3-I
models. They have been used for study of erosion of sand and mud in a
estuary and for study of 3-D movement of a plume of effluent in tida
waters.[43] An example is illustrated in Fig. 5 which shows the use of 2-L
nested models using progressively smaller meshes with a 3-D local mode
DAP can handle scalar variables, vectors and 64×64 matrices which ma
be real, integer or logical. The logical matrix is a very useful feature; it ca
be used to ensure that computations are only carried out at choser
locations, e.g. where there is water, not land, at a mesh point.
   At the mouths of most estuaries, channels and banks have convolutec
topographies which would need to be modelled using a fine rectangula
grid. That problem can be greatly reduced if a curvilinear grid can be use
with its margins parallel to local shorelines. There are several ways ir
which that can be done. Grid intersections do not necessarily have to b
orthogonal but, if they are, it may be possible to use potential flow theor
or at least a stream function to represent velocities. In that case it ha
proved possible to generate grids in which the curvature and spacing ar
a function of local depth.[49,63] Programmes have been written whicl

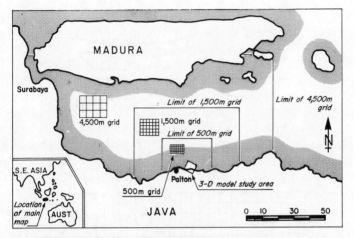

a) Location of Paiton Power Station

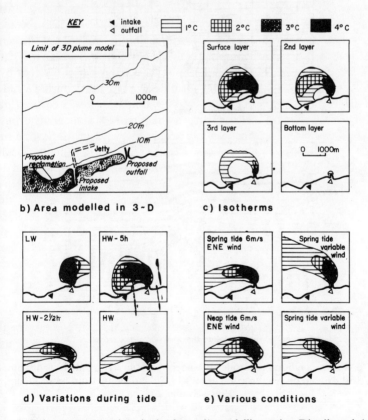

KEY ◀ intake ◁ outfall ☐ 1°C ☷ 2°C ▓ 3°C █ 4°C

b) Area modelled in 3-D

c) Isotherms

d) Variations during tide

e) Various conditions

FIG. 5. Paiton power station; hydrodynamic modelling using Distributed Array Processors. (Courtesy of Hydraulics Research Ltd.)

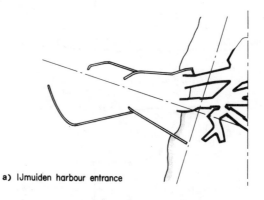

a) IJmuiden harbour entrance

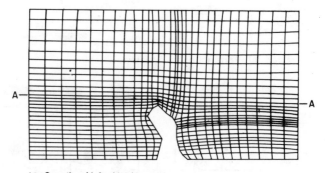

b) Smooth grid for IJmuiden area

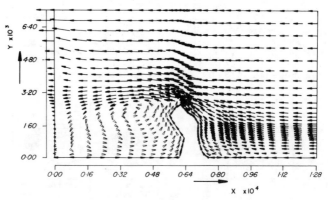

c) Velocity vectors using the $k - \varepsilon$ turbulence model

FIG. 6. Flow conditions outside Ijmuiden harbour, computed using model ODYSEE. (Based on reference 49, figures 19, 22 and 24, by courtesy of Pineridge Press, Swansea.)

generate curvilinear coordinates from depth contours. This has proved useful for study of small areas over which spatial changes of surface elevation can be neglected. Because the curvilinear grid is depth-generated, the depth cannot vary with time. This can be achieved in tidal situations by assuming a rigid lid as the upper boundary and specifying a free slip condition there. The effect of rise and fall of surface can be partially simulated by allowing pressures on the underside of the imaginary lid to vary. Figure 6 shows application of a rigid-lid model to Ijmuiden harbour entrance; Fig. 6(b) shows the computer-generated grid while Fig. 6(c) shows computed velocity vectors.

The choice of modelling method depends on the problems being studied. Numerical modelling specialists have a range of options available to them and combine them to achieve the best results at acceptable cost. It is possible to use a coarse rectangular grid for a regional model, and a stretched or curvilinear grid for local studies. The equations can be solved using fractional time steps as described in Section 6.[49] Implicit modelling methods have been used very widely, largely because of their inherent stability when using relatively large time steps. Their accuracy has been improved by various means. However, as more interacting phenomena are introduced, the equations become increasingly complicated and the advantages compared with explicit schemes diminish. Distributed array processors are very useful for complicated conditions. They can only be used with explicit methods, which are potentially more accurate than implicit when stability criteria are satisfied.

7 REGIONAL MODELLING OF WAVES

7.1 Modelling Methods

There are several well tried models which can be used to predict the 'global' wave climate in deep water at the boundaries of the region being studied.[31] They are beyond the scope of this chapter.

The primary objective of modelling waves is to determine the local climate of waves and currents so that sediment movements or dispersion of pollutants can be analysed. To this end, it is necessary to know the directional spectra of waves propagating into or through the region and to (i) analyse their modification by shallow water, friction and breaking, (ii) estimate the pattern of wave-induced currents, and (iii) analyse their interaction with tidal and river currents.

Solution of the full equations is difficult and would be prohibitively expensive if applied over an area typical of an outer estuary. Various compromises are necessary in order to reach engineering solutions. One is to divide the modelled area into regions in each of which different simplifying assumptions can be made. It is convenient to consider an outer and an inner region. In the outer region refraction is assumed to be the most important effect. Weak diffraction may also occur. The inner region includes refraction of waves by currents, all breaking waves and regions inshore of them, and thus includes the effects of radiation stresses and wave-induced currents. It is also the region of greatest rates of sediment transport. Some modelling methods bridge these regions but, for convenience, the outer region is discussed in this section while the inner region is considered in Section 8.

Four methods of modelling surface waves in transitional and shallow-water depths have been used.

(1) The most general approach is to obtain solutions to the wave equations of momentum and continuity in terms of surface elevation and particle velocity throughout the whole domain of interest at a particular time, and then to advance the solution in a series of time steps. This is necessarily extremely expensive in computer time, and simplifications must be made before it can be used even for small areas such as inside harbours. Practical economies can be made in suitable cases by using the Boussinesq equations, which apply to finite-amplitude waves with irrotational motion in inviscid water over a horizontal bed. Reasonable accuracy can be achieved at small values of depth/wavelength (< 0.15),[3] but the range of application can be increased by including higher-order terms in the difference equations.[2]

(2) If the wave climate changes only slowly with time, and linear theory is acceptable (small wave height), a steady-state harmonic solution for surface elevations and velocities can be assumed. The wave equation for this case is elliptical in form and must be treated as a boundary-value problem. It can include the effect of multiple reflections, refraction and diffraction. Possible solution methods are finite differences and finite elements. This model is only suitable for limited areas due to the high cost of computation, but requires less computing effort than solution of the Boussinesq equations.[30]

(3) Further simplification can be made if the effects of wave reflections are excluded from the solution. In that case, the equation can be approximated by one of parabolic form which can be solved as a problem of propagation, using a marching solution from the seaward boundary.[20] This results in considerable economies over elliptical-equation models.

These three modelling methods require computations to be done on a grid fine enough to resolve wave height over each wavelength—preferably at least 10 points, though fewer have been used. The grid spacing is typically 5–10 m.

(4) The fourth method is the ray method, based on geometrical optical theory. It is assumed that, in the absence of currents, waves propagate along orthogonals or rays and that energy is conserved between orthogonals. An initial-value problem can then be solved along each ray. The only inputs needed for monochromatic waves are period and water depth, and computation need only be done at intervals close enough to resolve significant changes in depth, typically of order 200 m. This method allows changes in concentration of energy flux to be calculated so that changes in relative wave height can be deduced using small-amplitude theory. The basic method does not allow diffraction or reflections to be taken into account.[59]

Each of these four general approaches has been developed so that different effects can be modelled. Variations of these methods have also been devised so that refraction of waves by depth and current changes can be studied.

7.2 Modelling Wave Refraction

The classic approach to study of wave propagation towards coasts was to apply simple refraction theory to discrete components of a wave spectrum and to build up as complete a picture as possible within available time and cost (e.g. Munk and Arthur[44]). Refraction occurs when wave propagation speed is changed locally as a consequence of changes in water depth. Small-amplitude wave propagation is governed by the linear dispersion equation $\omega^2 = gk \tanh kd$, from which wave propagation speed is $c = \omega/k$ and wave group velocity is

$$c_g = 0.5(1 + 2kd/\sinh(2kd))c$$

ω is angular wave frequency $= 2\pi/T$; T is wave period; k is wave number $= 2\pi/L$ where L is wavelength; d is mean water depth over a wave period. Wave rays can be determined for monochromatic waves of small amplitude. An approximation to real wave behaviour can be made by assuming that a spectrum consists of a number of discrete components that can be added linearly.

Bottom topography can cause wave rays to converge or diverge with resultant increase or decrease in energy flux per unit width. If rays converge to coincidence they form 'caustic' curves which indicate that theoretical energy flux per unit width becomes infinite. In reality, wave diffraction would occur and some wave energy would be dissipated by friction and breaking before reaching a predicted region of caustics.[4]

Modelling of simple refraction, e.g. by developing wave rays, is a very economical method because it does not require computation of wave profiles; it depends only on knowledge of local water depths. However, the assumption of linear superposition of discrete spectral components breaks down when steep finite waves occur and when waves approach regions of caustics. Development of wave rays is Lagrangian in form and is not a convenient method for computation of interaction of waves with tidal currents and of sediment transport.

The wave ray method can be adapted to an Eulerian form by using a finite difference method based on a fixed grid and computing wave properties at grid points by interpolation. By excluding diffraction from the analysis, wave phase and wave energy can be decoupled; this is, essentially, a feature of the wave ray method. Non-linear bottom friction and energy losses due to wave breaking can be included in such a model. Grid spacing is not critical to the analysis, the only requirement being that bottom topography should be adequately modelled. A grid spacing of 200 m or more is usually satisfactory. Using such a model, Sakai et al.[57] considered the propagation of spectral wave components from discrete directions at one fixed frequency. Booij et al.[12] developed a similar model to include changes in that frequency and associated changes in energy density. The alternative of choosing several frequencies, each with its energy concentrated in one particular direction, was developed by Chen and Wang.[14] A further development which includes diffraction, due to Yoo and O'Connor,[71] is described below.

7.3 Modelling Both Diffraction and Refraction

To deal with the problem of wave propagation over a rapidly varying topography, modelling of both refraction and diffraction is needed. Early models for diffraction were restricted to water of constant mean depth.[50] No refraction can occur under such conditions. Combined refraction and diffraction can be studied in models in which the wave equation is solved in two dimensions in plan.[68] The space-time grid must be fine enough to allow the wave profiles to be resolved adequately (5–10 m), yet the areas to be modelled must include extensive regions of shallow water. Such models are extremely expensive to operate, and their practical use is restricted to local regions where appreciable gradients of wave energy have been predicted by simple refraction models. In real situations, waves and currents have large random components which add to the complexity of analysis. Practical models must have limited objectives. There are many alternatives and the final choice will depend on the particular problem to alternatives and the final choice will depend on the particular problem to be solved. Models which can only provide steady-state solutions can be accepted if the real situation varies only slowly in its overall behaviour. Harmonic solutions may have to be accepted if their adoption results in very large savings in computer effort.

7.4 The 'Mild-slope' Equation

One compromise is to limit the model to gradually-varying bottom topography. The basis of this method is the 'mild-slope' equation first derived by Berkhoff.[8] It is elliptic in form. Valid solutions of this equation must be harmonic and steady-state.[9] Considerable economies in computation can be made if the equation can be modified to a parabolic form. This was done by Radder[54] and Booij[11] to include interaction with non-uniform currents (in plan) and energy dissipation effects due to wave breaking and bottom friction. Their version is valid for transient and non-harmonic solutions but it is only valid for progressive waves and cannot be used when there are strong wave reflections.

A full derivation of the mild-slope equation in various forms is given by Copeland.[15] It is based on conservation of wave energy, which requires that the rate of change of wave energy equals the rate at which work is done by external pressure:

$$\frac{DE}{Dt} + \iiint_\Gamma \mathbf{U} \cdot \Delta p \, dx \, dy \, dz = 0 \tag{11}$$

where E is wave energy per unit area; $\Delta = \partial/\partial x, \partial/\partial y, \partial/\partial z$; Γ represents a control volume.

Equation (11) leads to

$$\int_{-h}^{0} F\Delta^2\phi \, dz = 0 \tag{12}$$

where

$$F(z) = \frac{\cosh k(z + d)}{\cosh kd} \tag{13}$$

Two important conditions are imposed: (1) energy is conserved; (2) the bed must be mildly sloping, such that

$$|F\nabla^2\psi| \gg |\nabla F \nabla \psi| \tag{14}$$

where ψ is velocity potential at $z = 0$. The velocity potential

$$\phi(x, y, z, t) = \psi(x, y, t) \cdot F(z) \qquad (15)$$

Equation (14) leads to

$$kB \gg |\nabla B - g\nabla d/(\cosh^2 kd)| \qquad \text{(in general)} \qquad (16)$$

where $B = c \cdot c_g$, or

$$kc^2 \gg |\nabla c^2 - \nabla c^2/(\cosh^2 kd)| \qquad \text{(in shallow water)} \qquad (17)$$

which imply that there must not be appreciable changes in depth over the length of a wave.

In its steady-state, harmonic form the mild-slope equation due to Berkhoff[8] and modified by Smith and Spinks[60] and Lozano and Meyer[40] may be written

$$\nabla(B\nabla\psi) + k^2 B\psi = 0 \qquad (18)$$

This is an elliptic equation which can be solved as a boundary value problem. In its transient form without currents it may be written[11,15,16]

$$\nabla(B\nabla\psi) + (Bk^2 - \omega^2)\psi - \partial^2\psi/\partial t^2 = 0 \qquad (19)$$

Note that the mild-slope equation, in all its forms, is for monochromatic waves. The parabolic approximation only applies to waves with directions nearly collinear with one of the coordinate axes. Wave spectra and their modification cannot be reproduced directly, and the method breaks down if the waves undergo marked changes in direction as in the lee of islands, shoals or breakwaters. Parabolic solutions of the mild-slope equation have been based on the assumption of progressive waves. They do not apply to regions where there are strong wave reflections.

In order to overcome some of these deficiencies, Copeland[15,16] modified the mild-slope equation to a hyperbolic form. The steady-state harmonic form of the velocity potential is $\psi(x, y, t) = \phi(x, y)\exp(-i\omega t)$. From this,

$$\frac{\partial^2\psi}{\partial t^2} = -\omega^2\psi \qquad (20)$$

Inserting this in eqn. (18) leads to

$$\nabla(B\nabla\psi) - \frac{c_g}{c}\frac{\partial^2\psi}{\partial t^2} = 0 \qquad (21)$$

The surface condition $\psi = -i(g/\omega)\eta$ gives

This hyperbolic equation can be expressed as two first-order equations:

$$\nabla Q + \frac{c_g}{c} \frac{\partial \eta}{\partial t} = 0 \tag{23}$$

$$\frac{\partial Q}{\partial t} + B\nabla \eta = 0 \tag{24}$$

where Q is a dummy variable. It is harmonic in form, but its physical form is only needed on the model boundaries. $Q = c_g\eta$ is a solution provided that $\nabla c_g = 0$.

This is therefore useful in deep water or in water of constant depth. It can be calculated at model boundaries, provided only that depth at these boundaries is locally constant. Reflections can be included and, as there is no constraint on wave direction, it can be used down-wave of islands, shoals and breakwaters. The periodic solution of eqns (23) and (24) travels with phase speed given by $c^2 = (g/k)\tanh(kd)$, which implies conservation of wave action. There is no restriction on depth of water.

7.5 Analysis of Diffraction Using Wave Rays

Yoo and O'Connor[71] used a development of ray theory which has higher computational efficiency because it is based on wave dispersion rates rather than on detailed knowledge of wave profiles. The small-amplitude dispersion relationship is $\omega^2 = gk\tanh(kd)$. Battjes[4] had shown that, when diffraction occurs, the separation factor k should be replaced by wave number K:

$$K^2 = k^2 + \frac{1}{a} \frac{\partial^2 a}{\partial x_i^2} = k^2(1 + \delta) \tag{25}$$

where $i,j = 1,2$ (coordinate directions in plan), and a is wave amplitude.

Using this relationship in conjunction with the kinematic equation

$$\frac{\partial K_i}{\partial t} + \frac{\partial \omega}{\partial x_i} = 0 \tag{26}$$

Yoo and O'Connor obtained a wave crest conservation equation applicable to caustic or diffractive gravity waves:

$$\frac{\partial K_i}{\partial t} + K_j M \frac{\partial K_i}{\partial K_j} + S \frac{\partial d}{\partial x_i} - \frac{M}{2a} \frac{\partial^3 a}{\partial x_i \partial x_j^2} = 0 \tag{27}$$

where $M = 0.5(1 + G)(\omega/k)(1/k)$; $S = \omega G/(2d)$; $G = 2kd/(\sinh 2kd)$; d is total water depth $= h + \eta$; h is still water depth (including tidal rise and fall); η is mean surface elevation over a wave period.

The group velocity of caustic waves is

$$C_j = K_j M = 0.5(1 + \delta)(1 + G)(\omega/K)(K_j/K) \tag{28}$$

Equation (28) reduces to ordinary linear wave theory when $\delta = 0$.

The Battjes relation, eqn (25), also influences dynamic wave motion when diffraction occurs. Yoo[70] has shown that its effect is small and can be neglected when caustics are mild ($\delta \ll 1$). When dissipation effects are small, dynamic conservation of waves is given by

$$\frac{\partial a}{\partial t} + \frac{1}{2a} \frac{\partial}{\partial x_i} (K_i M a^2) = 0 \tag{29}$$

Yoo developed a numerical scheme for solution of these equations and compared its predictions with experimental results obtained by Ito and Tanimoto[30] for the case of a circular shoal which generated caustics when studied by simple ray theory. He also applied it to experimental results obtained by Berkhoff for a case of refraction and diffraction around twin ellipsoidal shoals, with good results.

8 MODELLING OF NEARSHORE WAVES AND CURRENTS

8.1 Introduction

At the sea-face of estuaries, wave action on terminal bars and sandbanks greatly enhances sediment transport due to tidal currents. In the major ebb- and flood-dominated channels, wave action at the bed is weaker but tidal currents are strong. There is a complicated pattern of sediment movement in the region, which varies according to the states of tide, wave action and fresh water discharge. Bed forms and, consequently, bed resistance undergo frequent change. At present there is little knowledge of bed forms caused by oscillatory flows combined with currents at an arbitrary angle though some measurements have been made in particular cases. This limits the potential accuracy of computation of local currents and wave action. Wave action causes radiation stresses which cause wave-induced currents, particularly near to and inshore of the breaker zone. In some cases, waves break on outer bars and re-form inshore of them. Knowledge of wave behaviour after breaking is still largely empirical. All these effects can be reproduced, but not in any one model. There is a range of models available, each of which can reproduce some aspects of the whole situation quite well. In breaking down a problem into several components, it has to be assumed that each behaves independently. This is not correct, and care has to be taken to ensure that the final result is a sufficient representation of reality for the purpose.

8.2 Wave/Current Interactions

In the nearshore region, wave breaking and wave-induced currents must be accounted for. Wave interactions with tidal currents can also be important.[26,32,33] To account for the effects of combined waves and currents, it is necessary to consider the motion of waves relative to the body of moving water as well as the absolute celerity of the waves. Calling the angular frequencies ω_r and ω_a respectively, the equations for interaction are

$$\mathbf{V} * \mathbf{K} = 0 \tag{30}$$

$$\frac{\partial \mathbf{K}}{\partial t} + \mathbf{V}\omega_a = 0 \tag{31}$$

$$\omega_a = \omega_r + \mathbf{K} \cdot \mathbf{U} \tag{32}$$

$$\omega_r^2 = gK \tanh(Kd) \tag{33}$$

where K is wave number, d is water depth, and U is depth-mean velocity. Symbols in bold type represent vectors, $*$ indicates vector product, and $\cdot$ scalar product. d includes the set-up or set-down of mean water level from tidal water level.

8.3 Solutions Based on the 'Mild-slope' Equation and Related Methods

One family of methods of modelling waves and currents is based on the mild-slope equation in its steady-state form, eqn (18), or its transient form, eqn (19). When tidal currents can be neglected, eqns (23) and (24) can be used. Once eqns (23) and (24) have been solved for a particular situation, radiation stresses can be calculated and used to estimate wave-induced currents which can be calculated with the aid of a steady-flow model. Copeland used this method to analyse wave set-up and wave-induced currents around an offshore breakwater in Liverpool Bay. In the form published by Copeland,[16] this model does not reproduce the effects of wave interaction with tidal currents.

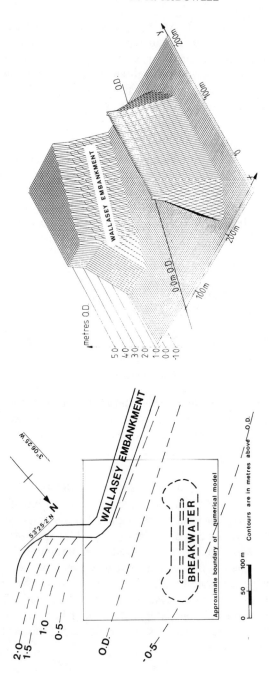

b) Discrete representation of bathymetry used in model

a) Plan of the Leasowe Bay breakwater

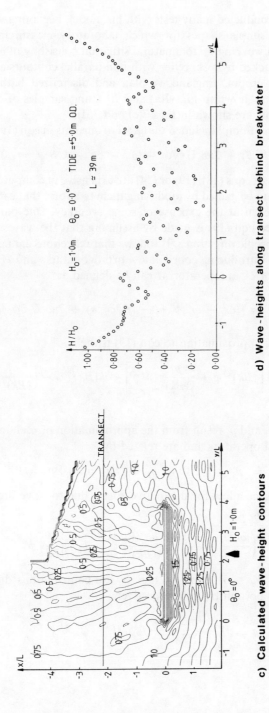

d) Wave-heights along transect behind breakwater

c) Calculated wave-height contours

Fig. 7. Wave conditions at Leasowe bay computed using Copeland's model. (Based on reference 15, figures 9.61 (a) and (b) and 9.62 (a) and (b), by courtesy of Dr G. Copeland.)

Copeland conducted many tests with his model, beginning with waves passing over simple shapes for which laboratory measurements were available, and working up to a natural situation consisting of an offshore breakwater backed by a shoreline with non-parallel contours at Leasowe Bay, near Liverpool, England. A plan and discretized bathymetry of Leasowe Bay is shown in Figs 7(a) and 7(b), and samples of calculations of wave heights are shown in Figs 7(c) and 7(d).

Radder[54] and Booij[11] included the effect of currents in eqn (19) to obtain

$$\nabla(B\nabla\psi) + 2i\omega_a\vec{U}\cdot\nabla\psi - (\omega_r^2 - \omega_a^2 - k^2B)\psi = 0 \qquad (34)$$

Full solutions of eqns (18), (19) or (34) are expensive in computer time and there is much to be gained by modifying them to a form that can be solved more easily, even at the expense of some accuracy. One parabolic approximation to eqn (19) is made by assuming that the waves propagate mainly in a specific direction s. It follows that reflections cannot be taken into account. Introducing coordinate n orthogonal to s and U_n and U_s as components of U, an operator M can be defined as[20]

$$M\phi = (\omega_a^2 - \omega_r^2)\phi + \frac{\partial}{\partial n}(B\partial\phi/\partial n) + 2i\omega_a U_n \partial\phi/\partial n \qquad (35)$$

The parabolic approximation to eqn (19) is then

$$\left(\frac{i\omega_a U_s}{B} + \frac{\partial}{\partial s}\right)\left[(Bk)^{0.5}\phi + \frac{p_1}{k(Bk)^{0.5}}M\phi\right] - ik(Bk)^{0.5}\phi - i\frac{p_2}{(Bk)^{0.5}}M\phi = 0 \tag{36}$$

Coefficients p_1 and p_2 result from the approximation of pseudo operators by differential operators and are related by

$$p_2 = p_1 + 1/2 \ (0 < p_1 < 1/2; \text{ optimal } p_1 = 1/4)$$

Additional terms allow for dissipative effects due to wave breaking and bottom friction:

$$\left(\frac{i\omega_a U_s}{B} + \frac{\partial}{\partial s}\right)p_1\frac{i\omega_a W}{k(Bk)^{0.5}}\phi + p_2\frac{\omega_a W}{(Bk)^{0.5}}\phi \qquad (37)$$

where $W = W_b + W_f$, and W_b and W_f are contributions due to wave breaking and bottom friction respectively.

$$W_f = \frac{S_0^2}{2g}\left(\frac{8}{3\pi}f_w S_0 H + 2f_{wc}U\right) \qquad (38)$$

where $S_0 = \omega_r/(\sinh(kd))$; f_w and f_{wc} are coefficients of wave-induced and wave-current-induced friction, taken as $f_w = 0.01$ and $f_{wc} = 0.005$.

Dingemans et al.[20] should be consulted for description of the model and its development tests. One test illustrates the potential of this model for study of sea conditions outside estuaries. It was used to calculate conditions in an area outside the Haringvliet sluices in the Rhine delta. Figure 8(a) shows contours and locations of measurements. Figure 8(b) shows computed iso-amplitude contours. Figure 8(c) shows samples of hindcast and measured wave-heights at a few locations. General agreement is good, but the model underestimates waves at station E-75 which is in the lee of a shoal. This demonstrates the inability of such models to cope with strong diffraction.

Solution of eqn (36) requires initial and boundary conditions. Initial conditions can be derived from the incoming wave field which can be

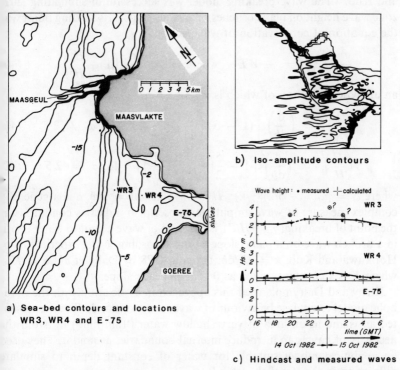

a) Sea-bed contours and locations WR3, WR4 and E-75

b) Iso-amplitude contours

c) Hindcast and measured waves

FIG. 8. Wave amplitudes at Haringvliet computed using parabolic approximation to mild-slope equation. (Based on reference 20, figures 14, 15 and 21, by courtesy of the American Society of Civil Engineers.)

weakly non-uniform in amplitude and direction. Lateral boundaries in a[
such numerical methods have to be provided for open boundaries alon[
which the wave field is not known. Dingemans et al. used

$$\cos\chi \, \frac{\partial\phi}{\partial s} + \sin\chi \, \frac{\partial\phi}{\partial n} \; = \; ik\phi \qquad (39$$

which ensures absorption of waves with local wave number k approachin[
at an angle χ to the direction of propagation, and partial absorption o[
waves from other angles. In the Dingemans et al. scheme, wave breakin[
was modelled according to a model of Battjes and Janssen.[5]

An interesting variation on this is due to Kirby and Dalrymple.[35] Thei[
parabolic-equation solution of the mild-slope equation was similar to tha[
of Dingemans et al., but their wave-breaking model was based on [
semi-empirical analysis by Dally et al.[19] and measurements by Horikaw[
and Kuo.[28] The wave-breaking model was successful in simulating surf
zone wave height on plane beaches of various slopes. Its starting point wa[
the equation of conservation of wave action:

$$\frac{\partial}{\partial\chi}\,(Ec_{\mathrm{g}}) \; = \; -\,WE \quad \text{where } W \; = \; \frac{\kappa c_{\mathrm{g}}}{d}\left(1 - \frac{H_s^2}{H^2}\right) \qquad (40$$

an analytical solution of which is

$$\left(\frac{H}{H_{\mathrm{b}}}\right)^2 \; = \; \left(\frac{\mathrm{d}}{\mathrm{d}b}\right)^2\left[(1-\Delta)\left(\frac{\mathrm{d}}{\mathrm{d}b}\right)^{\alpha-2\cdot5} + \Delta\right] \text{ for } \alpha \; \neq \; 2\cdot5 \qquad (41$$

$$\left(\frac{H}{H_{\mathrm{b}}}\right)^2 \; = \; \left(\frac{\mathrm{d}}{\mathrm{d}b}\right)^2\left[1 - \frac{5}{2}\left(\frac{\gamma}{\kappa}\right)^2\ln\left(\frac{\mathrm{d}}{\mathrm{d}b}\right)\right] \quad \text{ for } \alpha \; = \; 2\cdot5 \qquad (42$$

where $\Delta = [\alpha/(\alpha - 5/2)](\gamma/\kappa)^2$. H is wave height $= 2|a|$; a is a generall[
complex measure of wave amplitude; subscript b indicates conditions a[
the point of breaking; $\gamma = (H/d)_s$ is the ratio of wave height to water dept[
in a broken wave on a plane slope $= 0\cdot4$ according to the experiments o[
Horikawa and Kuo; $\kappa = H_{\mathrm{b}}/db$, taken as $0\cdot78$ by Dally et al.; $\alpha = K/s$
where K is a constant, taken as $0\cdot15$; s is beach slope.

Kirby and Dalrymple used this approach in association with the para
bolic model to analyse behaviour of waves around islands. They represen[
ted an island as an area of very shallow water ('thin film') to avoid th[
necessity of having to introduce internal boundaries around it. They use[
an analytical model suitable for water of constant depth to simulat[
diffraction in the lee of the island.

These parabolic approximations to the mild-slope equation cannot tak[
reflections into account. One approach to this has been developed b[

iu et al.[39] for the special case of water of constant depth. They obtained
olutions for wave motion past three-dimensional energy-dissipating
egions using a parabolic approximation with third-derivative terms. This
ould have value for modelling local regions such as breakwaters and
aining banks.

.4 Solutions Based on Wave Ray Models

nother family of models is based on the wave ray method. In the
resence of currents, wave rays and orthogonals are not coincident. It can
e assumed that wave action is conserved along rays, so that, if c_{ga} is the
bsolute wave group celerity and c_{gr} is the relative wave group celerity:

$$H^2 c_{gr} b / \omega_r = \text{constant}$$

here H is wave height and b is separation between wave rays.

Southgate[61] has shown that forward- and back-tracking methods based
n these equations give good predictions of the effects of wave/current
nteraction outside the breaker zone and in regions in which the bottom
opography does not change very rapidly. The effects of dissipation can be
aken into account by using a differential form of the wave action
quation:

$$\nabla \cdot (H^2 c_{ga} / \omega) = \text{dissipation rate}$$

here the right-hand side has a negative value.

A fixed-grid ray model based on refraction theory for linear waves has
lso been developed to reproduce spectral changes and refraction by
urrents.[29] Dissipation of wave energy occurs through bottom friction and
vave breaking. To estimate the former, water velocities at the bed due to
vaves must be known. The velocity spectral function based on linear
vaves was integrated over angle to give one-dimensional velocity spectral
unctions over frequency. Variances of these functions were then found by
ntegrating over frequency. The rate of energy loss by turbulent frictional
lissipation was described by

$$\frac{dE}{dt} = - P\rho |\bar{U}^3|$$

vhere P is a dimensionless friction factor, ρ is water density, and U is the
nstantaneous water particle velocity vector 'at the sea-bed'. The overbar
ndicates time-averaging over wave period. Because U is a two-dimension-
l random variable, it has a bivariate Gaussian distribution from which $\bar{U}^3$

can be calculated. Losses due to breaking waves were estimated by simple procedure designed to give a reasonable approximation to rea conditions. Wave heights were assumed to have a Rayleigh distributior and breaking was assumed to occur for wave heights in that distributio which exceeded $0.78d$. The total dissipation by breaking waves wa obtained by integrating the effect over all wave heights exceeding thi critical value. The paper included an example of computed spectral chang between offshore and inshore locations in a real situation. Figures 9(a) an 9(b) show changes for one inshore location, indicating that, at that poin refraction had caused local concentration of energy as well as change i spectral form. Figure 9(c) shows changes in root-mean-square wave heigh at twenty points close to the shore.

Yoo and O'Connor have also adapted their wave ray model fo combined waves and currents.[47,48] The result is a model which can re produce the effects of current and depth refraction, diffraction and energ dissipation due to bottom friction and breaking waves. It has all th economies of the wave ray method, being based on a grid which is onl required to reproduce bottom topography with sufficient accuracy However, it cannot reproduce the effect of reflected waves in its presen form. Their starting point was eqns (30)–(33), which are based on small amplitude theory, with eqn (25) to allow for diffraction of finite-amplitud waves. This led to a kinematic conservation equation for the wave numbe vector K_i.

$$\frac{\partial K_i}{\partial t} + (R_j + U_j)\frac{\partial K_i}{\partial x_j} + S\frac{\partial d}{\partial x_i} + K_j\frac{\partial U_j}{\partial x_i} - \frac{R}{2ka}\frac{\partial^3 a}{\partial x_i \partial x_j^2} = 0 \quad (43$$

where R_j are the intrinsic wave group velocity components given by

$$R_j = (K_j \omega_r)(1 + G)/(2k^2)$$

and

$$S = k\omega r/(\sinh 2kd)$$

$$G = 2kd/(\sinh 2kd)$$

subscripts $i,j = 1,2$; tensor notation corresponding to coordinate direc tions x,y.

Radiation stresses due to waves were described by

$$S_{ij} = [(1 + \delta)(1 + G)K_iK_j/K^2 + \delta_{ij}G_j](\rho g a^2/4) \quad (44$$

where $\delta_{ij} = 1$ for $i = j$ and 0 for $i \neq j$, while the wave-period and depth-averaged equations of motion were given as

$$\frac{\partial \eta}{\partial t} + \frac{\partial (dU_i)}{\partial \chi_i} = 0 \quad (45)$$

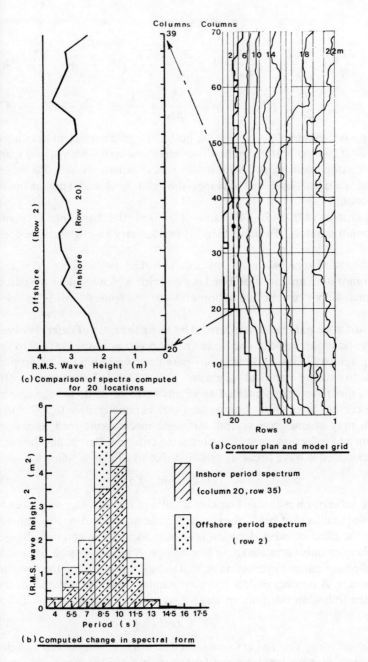

FIG. 9. Computed changes in wave spectra in a coastal region. (Based on reference 29, by courtesy of Hydraulics Research Ltd.)

$$\frac{\partial U_i}{\partial t} + U_j \frac{\partial U_i}{\partial x_j} + \frac{1}{\rho d} \frac{\partial S_{ij}}{\partial x_j} + g \frac{\partial \eta}{\partial x_i} + \frac{|U|}{d} C_c U_i = \frac{\partial}{\partial x_j} \varepsilon_j \frac{\partial U_j}{\partial x_j} \quad (46)$$

$$\frac{\partial a}{\partial t} + \frac{1}{2a} \frac{\partial}{\partial x_i} [(R_i + U_i)a^2] + \frac{S_{ij}}{\rho g a} + \frac{\partial U_j}{\partial x_i} + C_w a^2 = 0 \quad (47)$$

where ρ is the water density, C_c is the bottom friction coefficient associated with U_i, C_w is the bottom friction coefficient associated with a, and ε_j are eddy mixing coefficients. [Note that, in these equations, η is the wave-period averaged level and includes the tidal level and any set-up or set-down.]

Equations (43), (45), (46) and (47) form the basis of Yoo and O'Connor's model. To solve them, it is necessary to evaluate the coefficients C_c, C_w and ε_j.

In the surf zone, wave heights are controlled by breaking. Yoo and O'Connor used an improved breaking criterion which includes the effects of beach slope, current interaction and wave steepness. This is outlined below.

A surf zone parameter was formed by using the ratio of defect between kinematic group velocity and wave energy propagation velocity P to the group velocity. This theoretical surf parameter was found to be proportional to existing empirical parameters such as the Iribarren number. Its sound theoretical basis enabled an extension to be made for the case of wave/current interaction, leading to a surf parameter $\beta = (c_{gr} - P)/c_{ga}$ which may characterise several surf zone mechanisms such as set-up, run-up and reflection, as well as a breaking criterion. Battjes and Janssen[5] had extended a wave breaking criterion due to Miche as follows:

$$a_b = (\pi/7K)\tanh(q_b K d) \quad (48)$$

where subscript b indicates breaking point and the factor q_b was taken to include the effects of beach slope, current interaction and wave steepness. Since the effect of wave steepness is already included in eqn (45), Battjes and Janssen only correlated q_b to beach slope. It is now possible to include the effects of current interaction as well as beach slope, using the new surf parameter. A number of data sets was examined with the new parameter, and the following relationship was found to give satisfactory results:

$$q_b = 0.8 + \tanh(90\beta) \quad (49)$$

In the surf zone, Yoo and O'Connor used an eddy viscosity which included the effects of wave breaking and bottom friction of the combined wave/current flow. Full details are given in reference 70.

Bottom friction is one of the most important factors in balancing the forces due to radiation stresses in the nearshore circulation system. Several numerical models have been developed for determination of frictional stresses due to waves and currents, but for practical engineering calculations they are either too complicated or fall back on an assumption of constant friction factors over a whole domain. O'Connor and Yoo[47] adopted a simple method based on Bijker's approach[10] but modified to take into account the reduction in current velocity that occurs in the presence of wave action. Some details are given in Section 10 in the discussion of sediment transport.

An explicit method was used to solve the equations. It was considered that the need to solve simultaneous equations at each grid point removed the advantage of large Courant number that might have been possible with implicit methods.

There is no doubt that the methods developed by Yoo and O'Connor represent a major step forward in the search for numerical models that can give reliable answers over the large areas that have to be studied outside estuary mouths. One defect is their inability to include the effects of wave reflections in an interactive manner, which limits their use close to structures or steep shorelines.

9 SEA-BED STRESSES AND SEDIMENT TRANSPORT

One desirable goal of study of estuary interfaces is knowledge of sediment movement. The greatest rates of sediment transport occur in zones of breaking waves, which are found over offshore bars and shoals as well as along coastlines. The presence or absence of breakers depends on water depth and wave dynamics and will vary with tidal level for a given wave climate. Significant rates of transport can nevertheless occur elsewhere, particularly when long-period waves and tidal currents co-exist.

Equations for rates of sediment transport have been based on steady flows, and most have been linked to the depth mean velocity. This can be overcome if computations include two-dimensional velocity distributions over water depth, when it is possible to use estimates of shear stress at the bed to determine rates of pick-up of sediment and sophisticated turbulence models of the ε–κ type[56] to find concentrations of sediment in suspension. Variations of shear stress in space and time can then be used to estimate rates of erosion and deposition, e.g. in dredged trenches.[46,65] However,

these methods can only be used over limited areas and are not suitable fo
general study of estuary interfaces.

Wave action causes fine sediments to move at low current speeds, witl
the result that they are removed from most estuary interfaces, the residua
bed material being sand or coarse silt. For engineering purposes, estimate
of sediment transport over large areas have to be based on depth-averagec
velocities combined with estimates of wave action. It is fortunate that, ir
most cases, the resultant sediment transport can be regarded as bed load
If fine sediments occur in significant amounts, it is necessary to take intc
account the variation of flow velocities and turbulent intensities ove]
depth as a function of time. It should then be possible to use an approacl
such as Coles' law of the wake to describe velocity distributions in tida
and combined flows, but this has not yet been developed for practical use

When sediment moves mainly as bed load, a conservation equation car
be used to estimate changes in bed level:

$$\frac{\partial m}{\partial t} + \frac{\partial(T_s)}{\partial x} + \frac{\partial(T_s)}{\partial y} = 0 \tag{50}$$

where m (kg/m^2) is the mass of material on the bed, and T_s (kg/s/m width]
is the rate of transport of sediment. This only applies to bed load and fails
if there is a phase lag between rates of suspended and bed load. A specia]
situation occurs if part of the bed cannot be eroded or if local bed materia]
differs appreciably in its properties from the sediment in motion. Consider
for example, the co-existence of sand and shingle. Shingle moves much
more readily over a smooth sand bed than over a bed of shingle.
Moreover, shingle penetrates further through the boundary layer than
sand and is affected by much higher velocities and stresses than are sand
grains on a similar bed.

When sediment moves in suspension, the basic depth-averaged conservation equation, eqn (4), would apply, modified to take into account
the combined action of waves and currents. This equation depends on
specification of the source or sink term for sediment as well as on knowledge of initial and boundary conditions. Because there can be large
variations in sediment concentration over depth and with time, there are
inherent inaccuracies in depth-averaging. A more satisfactory approach is
to adopt a width-averaged model, for which a time-averaged conservation
equation can be written:

$$\frac{\partial}{\partial x}(bu\zeta) + \frac{\partial}{\partial z}[b(w - w_s)\zeta] - \frac{\partial}{\partial z}\left(b\varepsilon_{s,cw}\frac{\partial\zeta}{\partial x}\right) = 0 \tag{51}$$

where w is vertical velocity of flow, w_s is fall velocity of suspended sediment particles, b is flow width, ζ is sediment concentration at elevation z, and $\varepsilon_{s,cw}$ is turbulent mixing coefficient for sediment due to the combined action of waves and currents. Van Rijn[64] has described application of such a model to sedimentation in dredged channels by waves and currents.

To determine sediment transport rates or concentrations for use in these equations, it is necessary to determine shear stresses at the bed due to combined waves and currents. The general problem leads to complicated equations that do not lend themselves to use with the relatively simple depth-averaged calculations that are now possible. Simpler methods have been developed, the earliest being based on work done by Bijker.[10] Bijker assumed that the near-bed velocity profiles in currents and waves followed the logarithmic law based on simple mixing length theory. He then assumed that the near-bottom velocities due to waves and currents could be added vectorially. This, however, ignored interaction between currents and waves which is now known to be important. His basic method has been modified by others. Lundgren[41] and Grant and Madsen[22] assumed that the eddy viscosity due to wave action would be proportional to the maximum wave shear velocity at the bed. Lundgren proposed combination of shear velocity due to currents with a shear velocity based on the maximum wave velocity scaled to allow for average wave conditions. Grant and Madsen suggested that the eddy viscosity could be proportional to the maximum shear velocity of the combined flow. Both of these methods resulted in an improved representation of combined flow over the original Bijker method, but dependence on the maximum shear velocity rather than a wave-averaged value of some kind was a weakness. A further development was made by Fredsoe[21] who took into account the time-varying nature of boundary layer thickness and eddy viscosity. He showed that, under many real conditions, the bed was plane due to the intensity of wave action. For the case of a rough boundary, he assumed that the velocity profile was logarithmic within the wave boundary layer for ratios of wave velocity amplitude to bed roughness of over 30. The interaction of waves and currents was taken into account by assuming an increased value of effective bed roughness height when calculating current profiles. Vector addition of velocities due to waves and currents at the top of the wave boundary layer was then used to determine bed shear stresses, assuming a logarithmic velocity profile.

O'Connor and Yoo[47] used energy dissipation rates rather than shear stresses to describe an eddy viscosity of the combined flow. This made it easier to take into account interaction of the quasi-steady currents and wave action averaged over a wave period. The equation of motion of combined waves and currents, neglecting advective accelerations and horizontal diffusions, may be written

$$\frac{\partial u_w}{\partial t} + \frac{\nabla(p_c + p_w)}{\rho} = \frac{\partial}{\partial z}\varepsilon_a\frac{\partial u_c}{\partial z} + \frac{\partial}{\partial z}\varepsilon_a\frac{\partial u_w}{\partial z} \tag{52}$$

where ε_a is eddy viscosity due to combined flow over a wave period, u and p are velocity and pressure vectors respectively, and subscripts w and c refer to waves and currents respectively. If $\langle\ \rangle$ indicates average over a wave period,

$$\nabla p_c = \left\langle\frac{\partial}{\partial z}\varepsilon_a\frac{\partial u_c}{\partial z}\right\rangle + \left\langle\frac{\partial}{\partial z}\varepsilon_a\frac{\partial u_w}{\partial z}\right\rangle \tag{53}$$

If u_0 is the viscosity in the absence of waves, and the corresponding eddy viscosity is ε_c,

$$u_c = \alpha u_0 \text{ where, in this case, } \alpha = \varepsilon_c/\varepsilon_a$$

O'Connor and Yoo proposed using the energy dissipation rate to obtain the eddy viscosity for each case:

$$\varepsilon_m = \kappa^+ u_m^+ z \tag{54}$$

where

$$u_m^+ = (D_m/\rho)^{1/3} \tag{55}$$

m represents c, w and cw; κ^+ is a constant, analogous to von Karman's constant; u^+ is a scalar dissipation speed analogous to u^*, the vector shear velocity; D_m represents D_c, D_w and D_{cw}, the bottom energy dissipation rates for currents, waves and combined flows respectively.

Yoo[70] found a good correlation of eqn (54) with a constant value of κ^+, which he found to be about half von Karman's constant $\kappa = 0.4$.

The reduction factor for current velocity within the wave boundary layer is then

$$\alpha = (D_c/D_{cw})^{1/3} \tag{56}$$

O'Connor and Yoo derived the following equations which enable components of shear stress to be evaluated:

$$\langle\tau\rangle_p = \beta\tau_c \text{ parallel to the current} \tag{57}$$

$$\langle\tau\rangle_n = \gamma\tau_c \text{ normal to the current} \tag{58}$$

$$\beta = \alpha^2[1 + (0.36 - 0.14\cos 2\theta)(\mu/\alpha)^{1.5}] \tag{59}$$

$$\gamma = 0.205\alpha^{0.75}\mu^{1.25}\sin 2\theta \tag{60}$$

θ is the angle between the orbital motion and the normal to the current. The total frictional dissipation rate can be found from

$$D_{cw} = \beta\rho C_c U^3 + \delta\rho C_w \langle u_b \rangle^3 \tag{61}$$

where U is the depth-mean current velocity and $\langle u_b \rangle$ is scalar average velocity at the sea bed, with β from eqn (59) and δ from a numerically approximate analytical equation:

$$\delta = (4/3\pi) + (0.43 - 0.2\cos 2\theta)(\mu/\alpha)^{(0.1\cos 2\theta - 1.9)} \tag{62}$$

Equation (62) is an approximation to a rather complicated analytical expression based on mixing length theory.

Using eqn (61), the current reduction factor α is implicitly evaluated by

$$\alpha = (\beta + \delta C_w \langle u_b \rangle^3/(C_c U^3))^{-1/3} \tag{63}$$

O'Connor and Yoo used results of laboratory experiments with combined waves and currents to test these equations; details are given in reference 47. They found remarkably good agreement, with discrepancies between their predictions and measurements of less than 10% for 25 out of 26 comparisons, and less than 5% for 19 out of 26 comparisons. The only experimental results were for waves collinear with currents and at 90° to them. Nevertheless the theoretical basis of their equations gives reason to expect that they will also apply for any angle between waves and currents.

Once the shear stress at the bed has been determined, the rate and direction of bed-load transport can be estimated using any relationship that can be expressed in terms of shear stress rather than depth-mean velocity. This effectively rules out the stream-power relationships which have been so successful for steady flows, pending development of such relationships for combined waves and currents. Those based on the dimensionless particle parameter D_* and the transport stage parameter T can be used (see, for example, van Rijn[64]):

$$D_* = D_{50}[(s - 1)g/v^2]^{1/3} \tag{64}$$

$$T = [(U_*)^2 - (U_{*,cr})^2]/(U_{*,cr})^2 \tag{65}$$

where D_{50} is particle size, $s = \rho_s/\rho$, ρ_s is the density of solid particles, v is kinematic viscosity, $U_* = \tau_0/\rho$, the shear velocity at the bed, and $U_{*,cr}$ is the critical shear velocity at the start of bed movement.

10 CONCLUSIONS

Modelling of the sea at the mouths of estuaries is one of the most difficult of free-surface hydraulic engineering problems because of the great variety of forces at work. Problems can only be made manageable by treating most of them separately. Provided that the water is well mixed, tidal currents can be modelled using one of several well tried techniques. Recent developments have been in decoupling parts of the equations so that each can be solved using appropriate methods[6,7] and in utilising advanced techniques of computation such as Distributed Array Processors.[43] Developments in the study of dispersion in tidal waters have been described by Holly[27] and Sauvaget.[58] The main developments described in this chapter have been in study of waves in shallow water and their interaction with currents. The large areas that have to be modelled outside estuaries make representation of most hydraulic variables very demanding in computer time. All practical modelling methods are based on elimination of one or more of the effects such as reflection of waves, refraction and variation of wave climate with time. Real problems have to be solved using multiple models, each being appropriate to the combination of forces at work within its domain.

Recent developments have made it possible to use wave ray models based on a fixed grid to study the effects of refraction, diffraction, energy losses due to breaking waves and bottom friction and interaction with tidal currents. They cannot yet cope with wave reflections. It is now possible to use a relatively cheap modelling method based on a coarse grid —typically 200 m sides—over large areas in all cases where reflections can be neglected.

Table 2 is a comparison between examples of the main methods of modelling waves in tidal waters. It is not possible to give relative costs for their use since so much depends on the particular problem to be studied. However, they are placed in order of decreasing complexity. Those towards the left may be necessary for study of localised situations which may involve wave reflections and diffraction, e.g. close to sea-walls and dock entrances. Those towards the right are based on the wave ray method. They are the only ones suitable for study of large areas at reasonable cost.

This chapter began with a description of three quite different estuaries which posed problems representative of many others. Their main properties are summarised in Table 1. How could these estuaries be studied now?

TABLE 2
SUMMARY OF PROPERTIES OF WAVE MODELS

| Representative authors | Boussinesq type | Long wave | Mild slope and related | | | | Wave ray | | |
| | | | | | | | Nodal | | Ray |
	Abbott et al.[2,3]	Ito & Tanimoto[30]	Berkhoff[8]	Copeland[16,17]	Kirby & Dalrymple[35]	Booij et al.[12]	O'Connor & Yoo[48]	Hydraulic Research Ltd.[29]	Shore Protection Manual[59]
Time varying waves	√			√					
Harmonic waves	√	√	√	√	√	√	√	√	√
Refraction	√	√	√	√	√	√	√	√	√
Diffraction	√	√	√	√	Weak	Weak	√		
Reflection	√		√	√					
Scattering	√	√	√	√	Not directly[a]		√		
Interaction with currents		√				√	√	√	
Simulation of wave spectra								√	
Depth restriction		√							
Bed slope	None		Mild	Steep	Mild	Mild	Any	Any	Any
Grid resolution: Wavelength (< 10 m)	√	√	√	√	√	√			
Bottom contours (> 100 m)							√	√	√
Number of computational operations for M nodal grid points		M^3	M^4	M^3	M^2				

[a] See reference 34 for recent developments.

The Santos estuary (Fig. 1) has a small tidal range. The dominant wave action is a swell approaching from south-east with a typical mean period of about 11 s, but passage of cold fronts causes waves to be generated locally and also causes variations in sea-level which are comparable to the tidal range. The rocky coast on either side of the bay causes local reflection of waves but the effect is likely to be small near the centre of the bay where the main engineering problems occur. They are the approach channel to the estuary and port and a major sewerage outfall.

Tidal currents are weak but have an important effect on the direction of sediment transport which is initiated by waves. Because the bay is partially mixed, numerical modelling would have to differentiate between currents close to the bed and depth-mean currents. The simplest model that would have any possibility of reproducing this effect would be a two-layered two-dimensional plan model; such models are available in major hydraulics institutions. However, it would be better to use field measurements for this; they would be needed in any case for calibration of any models used. As it happens, extensive field measurements have established the patterns of circulation in the bay in sufficient detail for initial engineering design studies.

Waves are modified by refraction and enter the bay over a rather narrow arc. Swell could be represented well by monochromatic waves. The locally generated waves are more variable but their frequency and intensity may not justify more sophisticated modelling. Reflections from cliffs will have a local effect which would have to be explored further, but they are not likely to make much difference to sediment or water movement in the centre of the bay or in the main navigation channel. Because tidal currents are weak—less than 0·5 m/s except at the mouth of the estuary—they will not cause appreciable refraction of waves. However, waves at the shore induce alongshore movement of sediment which gives rise to a circulation of sediment into the navigation channel.

Numerical models could consist of the simplest possible nodal wave ray models, one of which would enable waves near the centre of the bay to be forecast.[29] From that point, where the effect of refraction may cause local caustics to develop, greater accuracy would be achieved by reproduction of diffraction. The O'Connor and Yoo model[48] would be suitable for this region. This model would also be used for estimation of resultant shear stresses at the bed and sediment transport. Sedimentation in the navigation channel would best be estimated by use of a width-averaged 2-D model.[65,46]

The River Pungue (Mozambique) (Fig. 2) has quite strong semi-diurnal tides, extensive offshore shoals and channels and moderate wave action. There is a current flowing from north to south (the equatorial current) parallel with the east coast of Africa, but giving rise to large eddies in coastal embayments including that of the mouth of the Pungue. As at Santos, the dominant waves approach from the antarctic through a rather narrow sector, centred on south-east. It is essential to model the combination of tidal currents, waves breaking on the sandbanks and the resultant wave-induced currents. A physical model has been built representing the estuary and adjacent sea but, because of scale distortion, it was not able to reproduce the effect of breaking waves and wave-induced circulations. Field measurements have limited use due to the large area to be studied and the difficulty of measuring in most of the regions where sediment movement is taking place.

It is essential to model the current field, including the tidal and equatorial currents. For this purpose the modelling system developed by Benque et al.[6,7] could be used. Alternatively, it would be possible to use a model based on Distributed Array Processors[43] which would allow 'patching' of regions where rapid changes of topography occur, using a finer computational grid.

Modelling of waves would also be done in a suite of models. In an outer zone, the simplest possible wave-ray model could be used.[59,29] Near the sandbanks, caustics would develop and simple ray models would break down. The O'Connor and Yoo model[47,48] could be used here, with a grid size of about 200 m. It would be able to represent the modification of monochromatic waves by breaking and bottom friction as they move to the shore. Close to the shore itself, and particularly near high water level when coast-protection walls and quays would reflect waves, it might be necessary to use a model that can handle reflections, such as Copeland's hyperbolic model.[16,17] However, that can only be used when tidal currents are weak. It might be necessary to use a model based on the long-wave equations such as that of Ito and Tanimoto.[30] These models could only be used very close to the shore because the grid size would have to be of order 6 m.

The most important feature of the outer Pungue estuary is the circulation of water and sediment by wave-induced currents and tides from the outer shoals towards the shore and then along the shore towards the mouth of the estuary. From there, ebb tides return much of the sediment to the outer sandbanks. It is evident that, in this case, recently developed methods can model effects that cannot be reproduced by any other means.

The Mersey estuary (Fig. 3) has large tidal rise and fall, very strong currents and high input of wave energy. It lacks the long-period swell which is characteristic of the other two estuaries described. Waves can approach Liverpool Bay from a sector of width about 90° centred on north-west. The mouth of the estuary is flanked by vertical walls, as is the coast south of the entrance. There are extensive offshore sandbanks through which the main ebb channel carries ships between training revetments. These revetments are made from random rubble with crests 1–2 m above lowest spring tide level. Engineering studies are likely to be concerned with sedimentation in the navigation channel, circulation of material dumped in Liverpool Bay, local coastal erosion and studies in connection with a proposed barrage across the estuary.

Tidal currents have a considerable effect on wave refraction. Wave spectra are modified by waves breaking on outer shoals and by bottom friction. When water levels rise above mean tide level, waves penetrate to the shore at all points and reflections from sea walls occur. These become a major factor near high tide level. All aspects of modelling waves and currents are significant in this case. When tide levels are near low water, wave action can only reach the outer margins of the shoals and the current is confined to Liverpool Bay and the main channel. Modelling is not likely to be needed then. For water levels between mean and high tide, a modelling method that can reproduce refraction, diffraction, interaction with currents and dissipation by wave breaking and bottom friction is needed. The O'Connor and Yoo[47,48] model is capable of doing this over most of the area to be studied, but it cannot represent reflections from walls near the entrance to the estuary which become important when levels approach high tide. It has the advantage of only requiring a coarse grid, of say 200 m side. Near to the shore, and particularly when tidal currents are weak, the Copeland[15-17] or Ito and Tanimoto[30] models can be used. They would require a grid of 5 m or 6 m side to represent the rather short waves present in Liverpool Bay.

Developments in the whole field of sediment transport due to waves and currents are taking place rapidly. The models described here are being refined and developed for commercial use. Subjects in need of development at the moment are: representation of wave spectra in the presence of currents and their modification as they lose energy through breaking and bottom friction; the mechanics of sediment transport due to combined waves and currents; representation of unsteady wave climates; effective methods of monitoring sediment transport close to the bed. Work is being done on all these topics and clear progress with them should have been reported by 1992. Associated with these developments is the need for well coordinated field measurements so that they can be properly evaluated.

REFERENCES

1. ABBOTT, M. B. *Computational Hydraulics*, Pitman, London, 1979.
2. ABBOTT, M. B., McCOWAN, A. D. and WARREN, I. R. Accuracy of short-wave numerical models, *J. Hydraulic Engng ASCE*, **110**(10) (1984), 1287–1301.

3. ABBOTT, M. B., PETERSEN, H. M. and SKOVGAARD, O. On the numerical modelling of short waves in shallow water, *J. Hydraulic Res.*, **16** (1978) 173–204.

4. BATTJES, J. A. Refraction of water waves, *Proc. ASCE*, **94**(WW4) (1968), 437–51.

5. BATTJES, J. A. and JANSSEN, J. P. F. M. Energy loss and set-up due to breaking of random waves, *Proc. 16th Conf. Coastal Engng, ASCE*, 1978, pp. 569–89.

6. BENQUE, J. P., CUNGE, J. A., FEUILLET, J. and HOLLY, F. M. New method for tidal current computation, *Proc. ASCE*, **108**(WW3) (1982), 396–417.

7. BENQUE, J. P., HAUGUEL, A. and VIOLLET, P. L. *Engineering Applications of Computational Hydraulic Models*, Vol. II, Pitman, London, 1982.

8. BERKHOFF, J. C. W. Computation of combined refraction diffraction, *Proc. 13th Conf. Coastal Engng*, ASCE, 1972, pp. 471–90.

9. BERKHOFF, J. C. W., BOOIJ, N. and RADDER, A. C. Verification of numerical wave propagation models for simple harmonic water waves, *Coastal Engng*, **6** (1982), 255–79.

10. BIJKER, E. W. The increase of bed shear in a current due to wave action, *Proc. 10th Conf. Coastal Engng*, ASCE, **1** (1966), 746–65.

11. BOOIJ, N. Gravity waves on water with non-uniform depth and current, Dept. of Civil Engng, Delft University of Technology, Report No. 81-1, 1981.

12. BOOIJ, N., HOLTHUIJSEN, L. H. and HERBERS, T. H. C. A numerical model for wave boundary conditions in port design, Int. Conf. Numerical and Hydraulic Modelling of Ports and Harbours, BHRA, Birmingham, England, 1985.

13. BREBBIA, C. E. M. FEM in design of hydraulic structures. In: *Developments in Hydraulic Engineering*, Vol. 1, ed. P. Novak, Elsevier Applied Science Publishers, London, 1983.

14. CHEN, Y. and WANG, H. A numerical model for non-stationary shallow-water wave spectral transformations, *J. Geophys. Res.*, **88**(C14) (1983), 9851–63.

15. COPELAND, G. J. M. A numerical model for the propagation of short gravity waves and the resulting circulation around nearshore structures, PhD thesis, Liverpool University, 1985.

16. COPELAND, G. J. M. A practical alternative to the 'mild-slope' equation, *Coastal Engng*, **9** (1985), 125–49.

17. COPELAND, G. J. M. Practical radiation stress calculations connected with equations of wave propagation, *Coastal Engng*, **9** (1985), 195–219.

18. CRICKMORE, M. J. and SHEPHERD, I. E. A field instrument for measuring the concentration and size of fine sand suspensions, Int. Conf. Measuring Techniques of Hydraulics Phenomena in Offshore, Coastal and Inland Waters, BHRA, London, 1986, Paper K2.

19. DALLY, W. R., DEAN, R. G. and DALRYMPLE, R. A. Wave height transformations across beaches of arbitrary profile, *J. Geophys. Res.*, **90**(C6) (1985), II. 917–27.

20. DINGEMANS, M. W., STIVE, M. J. F., KUIK, A. J., RADDER, A. C. and BOOIJ, N. Field and laboratory verification of the wave propagation model CREDIZ, *Proc. 19th Conf. Coastal Engng*, ASCE, September 1984.

21. FREDSOE, J. Turbulent boundary layer in wave-current motion, *J. Hydraulic Engng ASCE*, **110**(8) (1984), 1103–20.

22. GRANT, W. D. and MADSEN, O. S. Combined wave and current interactio
    with a rough bottom, *J. Geophys. Res.*, **84**(C4) (1979), 1797–1808.
23. HEAPS, N. S. Density current in a two-layered coastal system with applicatio
    to the Norwegian coastal current, *Geophys. J. Roy. Astron, Soc.*, **63** (1980)
    289–310.
24. HEAPS, N. S. Three-dimensional modelling for tides and surges with vertica
    eddy viscosity prescribed in two layers, Parts I and II, *Geophys. J. Roy. Astron
    Soc.*, **64** (1981), 292–310.
25. HEATHERSHAW, A. D. and CARR, A. P. Measurements of sediment transpor
    rates using radioactive tracers, *Proc. Coastal Sediments Conf.*, ASCE, 1977.
26. HEDGES, T. S. Some effects of currents on wave spectra, *Proc. 1st Indian Conf
    Ocean Engng*, I.I.T., Madras, 1981.
27. HOLLY, F. M. Dispersion in rivers and coastal waters. 1. Physical principle
    and dispersion equations. In: *Developments in Hydraulic Engineering*, Vol. 3
    ed. P. Novak, Elsevier Applied Science Publishers, London, 1985.
28. HORIKAWA, K. and KUO, C.-T. A study of wave transformation inside
    surfzone, *Proc. 10th Conf. Coastal Engng*, Tokyo, Japan, 1966, pp. 217–233
29. HYDRAULICS RESEARCH LTD. A finite difference wave refraction model, Repor
    No. EX 1163, April 1984.
30. ITO, Y. and TANIMOTO, K. A method of numerical analysis of wave propaga-
    tion: application of wave refraction and diffraction, *Proc. 13th Conf. Coasta
    Engng*, ASCE, 1972, Chapter 26.
31. JANSSEN, P. A. E. M., KOMEN, G. J. and DEVOOGT, W. J. P. An operational
    coupled hybrid wave prediction model, *J. Geophys. Res.*, **8**(C3) (1984), 3635–
    54.
32. JONSSON, I. G. Combinations of waves and currents. In: *Stability of Tida
    Inlets*, ed. P. Brunn, Elsevier, Amsterdam, 1978, pp. 162–203.
33. JONSSON, I. G. and WANG, J. D. Current-depth refraction of water waves.
    Ocean Engng, **7** (1980) 153–71.
34. KIRBY, J. T. Rational approximations in the parabolic equation method for
    water waves, *Coastal Engng*, **10**(4) (1986), 355–78.
35. KIRBY, J. T. and DALRYMPLE, R. A. Modelling waves in surfzones and around
    islands, *J. Waterway, Port, Coastal and Ocean Engng ASCE*, **112**(1) (1986),
    78–94.
36. KOUTITAS, C. G. *Elements of Computational Hydraulics*, Pentech Press,
    London, 1983.
37. KOUTITAS, C. G. and O'CONNOR, B. A. Modelling three-dimensional wind
    induced flows, *J. Hydraulics Div. ASCE*, **106**(11) (1980), 1843–65.
38. LEENDERTSE, J. and LIU, S. 3-D flow simulation in estuaries, *Proc. Int. Symp.
    Unsteady Flow in Open Channels*, BHRA, England, 1976.
39. LIU, P. L.-F., YOON, S. B. and DALRYMPLE, R. A. Wave reflection from energy
    dissipation region, *J. Waterway, Port, Coastal and Ocean Engng Div. ASCE*,
    112(6) (1986), 632–44.
40. LOZANO, C. and MEYER, R. E. Leakage response of waves trapped by round
    islands, *Physics of Fluids*, **19**(8) (1976), 1075–88.
41. LUNDGREN, H. Turbulent currents in the presence of waves, *Proc. 13th Conf.
    Coastal Engng ASCE*, 1972, 623–34.

42. McDOWELL, D. M. and O'CONNOR, B. A. *Hydraulic Behaviour of Estuaries*, Macmillan, London, 1977.

43. MILES, G. V. and COOPER, A. J. Application of a DAP computer to tidal problems, *Int. Conf. Numerical and Hydraulic Modelling of Ports and Harbours*, BHRA, Birmingham, England, 1985.

44. MUNK, W. H. and ARTHUR, R. S. Wave intensity along a refracted bay, Symp. on Gravity Waves, National Bureau of Standards, Washington, 1952, circular No. 521.

45. NOVAK, P. and CABELKA, J. *Models in Hydraulic Engineering*, Pitman, London, 1981.

46. O'CONNOR, B. A. and TUXFORD, C. Modelling siltation at dock entrances, 3rd Int. Symp. Dredging Technology, BHRA, 1980, Paper F2.

47. O'CONNOR, B. A. and YOO, D. Mean bed friction of combined wave-current flow, *Coastal Engineering* (1988) (in press).

48. O'CONNOR, B. A. and YOO, D. Mathematical modelling of wave-induced nearshore circulations, *Proc. 20th Conf. Coastal Engng, ASCE*, Taipei, Taiwan, 1986.

49. OFFICIER, M. J., VREUGDENHIL, C. B. and WIND, H. G. Applications in hydraulics of numerical solutions of the Navier–Stokes equations. In: *Recent Advances in Numerical Fluid Dynamics*, ed. C. Taylor, Pineridge Press, Swansea, 1984.

50. PENNEY, W. G. and PRICE, A. T. The diffraction theory of sea waves and shelter afforded by breakwaters, *Phil. Trans. Roy. Soc.*, **244A** (1952), 236–53.

51. PILLSBURY, G. B. *Tidal Hydraulics*, Ch. VII, Corps of Engineers, US Army, 1956.

52. PRANDLE, D. A numerical model of the southern North Sea and River Thames, Report No. 4, Institute of Oceanographic Sciences, 1974.

53. PRANDLE, D. and HOWARTH, J. The use of HF radar measurements of surface currents for coastal engineers, Int. Conf. Measuring Techniques of Hydraulics Phenomena in Offshore, Coastal and Inland Waters, BHRA, London, 1986, Paper A1.

54. RADDER, A. C. On the parabolic equation method for water wave propagation, *J. Fluid Mechanics*, **95**(I) (1979), 159–76.

55. RODENHUIS, G. S., BRINK-KJOER, O. and BERTELSEN, J. A. A North Sea model for detailed current and water level predictions, *J. Petroleum Technol.* (1978), 1369–76.

56. RODI, W. Turbulence models and their application in hydraulics, published for IAHR, by A. A. Balkema, Rotterdam, 1984.

57. SAKAI, T., KOSEKI, M. and IWAGAKI, Y. Irregular wave refraction due to current, *Proc. ASCE*, **109**(9) (1983), 1203–15.

58. SAUVAGET, P. Dispersion in rivers and coastal waters. 2. Numerical computational of dispersion. In: *Developments in Hydraulic Engineering*, Vol. 3, ed. P. Novak, Elsevier Applied Science Publishers, London, 1983.

59. SHORE PROTECTION MANUAL, US Army Coastal Engineering Research Center, Corps of Engineers, USA, 1984, Chapter 2.3.

60. SMITH, R. and SPINKS, T. Scattering of surface waves by a conical island, *J. Fluid Mechanics*, **72** (1975), 373–84.

61. SOUTHGATE, H. N. Current-depth refraction of water waves, Hydraulic Research Ltd., Report No. SR14, January 1985.
62. STELLING, G. S., WIERSMA, A. K. and WILLEMSE, J. B. T. M. Practical aspect of accurate tidal computations, *J. Hydraulic Engng ASCE*, **112**(9) (1986) 802–17.
63. THOMPSON, J. F., WARSI, Z. U. A. and MASTIN, C. W. Boundary fitte coordinate systems for numerical solutions of partial differential equations: review, *J. Comp. Physics*, **47**(1) (1982), 1–108.
64. VAN RIJN, L. C. Sediment transport. Part I. Bed load transport, *J. Hydrauli Engng ASCE*, **110**(10) (1984), 1431–56.
65. VAN RIJN, L. C. Sedimentation in dredged channels by currents and waves, *J Waterway, Port, Coastal and Ocean Engng ASCE*, **112**(5) (1986), 541–9.
66. VINCENT, C. E., HANES, D., TAMURA, T. and CLARKE, T. L. The acousti measurement of suspended sand in the surf zone, Int. Conf. Measurin Techniques of Hydraulics Phenomena in Offshore, Coastal and Inland Waters, BHRA, London, 1986, Paper K3.
67. WEARE, T. J. Errors arising from irregular boundaries in ADI solutions of the shallow water equations, *Int. J. Numerical Methods Engng*, **14** (1979), 921–31
68. WILLIAMS, R. G., DARBYSHIRE, J. and HOLMES, P. Wave refraction and diffrac tion in a caustic region: numerical solution and experimental verification *Proc. Inst. Civil Engrs*, Part 2, **69** (1980), 635–49.
69. WYATT, L. R., VENN, J. and BURROWS, G. D. Ocean surface wave and curren measurements with HF ground-wave radar, Int. Conf. Measuring Techniques of Hydraulics Phenomena in Offshore, Coastal and Inland Waters, BHRA London, 1986, Paper A2.
70. YOO, D. Mathematical modelling of wave current interacted flow in shallow waters, PhD thesis, University of Manchester, 1986.
71. YOO, D. and O'CONNOR, B. A. A ray model for caustic gravity waves, Proc 5th Congr. Asian and Pacific Regional Division, IAHR, August 1986.

Chapter 4

POLDERS

J. Luijendijk,[a] E. Schultz[b] and W. A. Segeren[a]

[a] *International Institute for Hydraulic and Environmental Engineering (IHE), Delft, The Netherlands*

[b] *IJsselmeerpolders Development Authority, Lelystad, The Netherlands*

1 INTRODUCTION

The cultivated area on earth is about 15 million km^2. From this area about 2·7 million km^2 are irrigated, and about 1·5 million km^2 drained.[1] A special type of drained land is a 'polder', generally found in low-lying coastal areas, river plains, shallow lakes, lagoons and upland depressions.

For the reclamation of polders several aspects have to be taken into account, especially the technical aspects dealing with flood protection (dikes), drainage, and possibly irrigation, but also social, economical and environmental aspects which may be crucial for the success of polder development.

In this chapter the emphasis will be on the technical aspects of polder development.

2 DESCRIPTION OF POLDERS

Since far back in history man has been trying to make use of, and protect himself against, water. Although the word polder did not exist at that time, works were executed that could be called polder development. The first recorded activity dates from the fourth millennium BC, when the Sumerians reclaimed the marshy low-lying areas of Mesopotamia. But flood protection, irrigation and drainage works which can be considered as impoldering[2] were also carried out in Egypt, in the valleys of the Indus and the Ganges, in China, Peru and Central America.

The word polder originates from a country that is often identified with that word: the Low Lands, the Netherlands, Les Pays-Bas, Los Paises Bajos.

Before impoldering, polder areas were either waterlogged or temporarily or permanently under water. In general this meant that for at least a period of time during the year the amount of incoming water dominated the discharge from the area in such a way that the ground water level and/or open water level rose so much that agricultural activities became impossible. The incoming water can originate from different sources (see Fig. 1):

— the sea: once or twice a day during high tide; a few days per month during spring tide; occasionally during storm surge; permanently (shallow lagoons)
— a river or a canal: once or twice a day in a tidal river; seasonally during high discharges
— a lake: mostly seasonally, or permanently
— seepage: mostly permanently from surrounding areas with a higher water level
— rainfall: frequently; seasonally (monsoon, wet winter period)

Recently several definitions for a polder have been given. The most

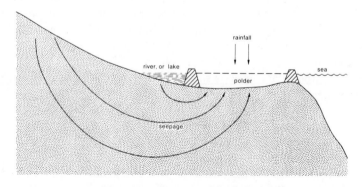

FIG. 1. Sources of water surplus.

widely used ones are:

— A tract of low land reclaimed from the sea, or other body of water, by dikes, etc. In the polder, the runoff is controlled by sluicing or pumping, and the water table is independent of the water table in the adjacent areas.[3]

— A level area which was originally subjected to a high water level, either permanently or seasonally due to either ground water or surface water. It becomes a polder when it is separated from the surrounding hydrological regime in such a way that its water level can be controlled independently of its surrounding regime.[4]

Although the last definition also includes a paddy-field on high terraces, or a flood plain of a large river protected by an upstream dam, none of these areas is ever called a polder.

This chapter will be based on the last definition of a polder which leaves room for different configurations and stages of development. Examples of this can be obtained from the history of polder development in the Netherlands. It was a slow irregular process that took about a millennium to reach the more or less final stage, starting with the construction of small dikes and simple structures, still making use of gravity for drainage. An important step forward could be taken after the introduction of the windmill, followed by steam-powered pumping stations some centuries later leading to independence from the natural elements. Since the last century electrical and diesel power has been widely applied.

The whole development process from simple intervention in the natural hydrological system to an advanced polder with optimal water management is determined by social, economic and environmental changes together with the expansion of technical possibilities.

At what particular moment an area can be regarded as a polder is not always clear, because one has to know when the water level could be controlled independently of the surrounding regime. It is really a matter of the quality of the water management system, because full independence never exists. In this contribution all the polders from their initial development stage to the ultimate advanced stage will be included in the definition.

Polders can be found all over the world, in different climates, below mean sea level but also far above, in different forms and qualities. As far as we know, up to the present, about $300\,000\,\text{km}^2$ throughout the world have been converted into some sort of polder; see Table 1.

From the definition, distinction can be made between polders in areas

TABLE 1

IDENTIFIED AREAS OF POLDERS IN THE WORLD

Continent	Polder area (km^2)
Africa	20 880
Asia	159 290
Australia	150
Central America and the Caribbean	20
Europe	64 600
North America	6 800
South America	825
Total	252 565

where a regional drainage base exists either permanently (swamps, shallow sea and lake beds) or temporarily (tidal lowlands, seasonally flooded river plains).

Apart from flood protection, the basis for the good functioning of a polder is a good water management system. When the outer water level is permanently above the desired inner level, the latter can only be maintained by pumping the excess water out of the polder area. Where the regional drainage base is controlled by the sea or a river, excess water may be discharged from the polder by gravity during periods of low tide, or during low river flow periods when the outer level falls below the inner level.

Based on the above, three different groups of polders can be distinguished:

— impoldered low-lying lands
— lands reclaimed from the sea
— drained lakes.

There are several characteristic aspects in a description of polders, particularly their level of protection, water management system and land use.

In determining the *level of protection* the values within the protected area as well as outside conditions play a role. Within the protected area a major distinction has to be made in value of property and human life.

Protection of property can be approached empirically, or based on a cost benefit analysis.[5] If floods can cause loss of human life, then a very high level of safety has to be chosen for the design of protection works or, for example, artificial mounds have to be constructed to which people can flee in an emergency.

In relation to outside conditions, it is important whether the sea, a river, lake, or canal borders the polder. Although differences occur within each of these groups, some general characteristics are important, such as the behaviour of floods, the possibilities of forecasting and the influence of the wind.

Most dangerous flooding is normally caused by the sea. This generally has the disadvantage that only short-term forecasting of some hours can be made and that the wave action can be very destructive. Floods from rivers can nowadays often be predicted some days beforehand. This, and the reduced wave action, may result in a decision for a lower level of security for polders along rivers than for polders along the sea. If polders bordering lakes or canals are flooded, then the flood is normally caused by a catastrophe and not by a hydrological extreme. In most cases the water body causing the flood as well as the areas that can be influenced will be small.

The *water management system* consists of a drainage system that can be combined with an irrigation system. The drainage system can consist of a land drainage system, or tertiary system to drain the soil, a hydraulic transport system, or primary (secondary) system to transport the water from the tertiaries to the outlet and an outlet structure to evacuate the water from the area.

A typical aspect of the drainage system is that it only has to drain the surplus rainwater and seepage water that reaches the area and not surface water from surrounding areas.

Another subdivision of the drainage system may be based on the discharge structure, i.e. sluices, draining by gravity or pumping stations. In areas with high intensity rainfalls, high pumping capacities may be required but often only for short periods.

The irrigation system may consist of a field irrigation system, a main system and an inlet structure. Low-lying polders may have the advantage of a relatively high outside water table also in dry periods. The inlet of irrigation water can thus be easily established.

As far as *land use* is concerned most of the existing polders are created for agricultural use. Dry food crops and rice must be distinguished, but mainly small polders have been made for urban or industrial use. Special examples of multiple land use from the very beginning of development are the Flevoland polders in the Netherlands (see Fig. 2), and Hachirogata in Japan. Another land use aspect in polders is that, when densely populated areas surround the polder, several decades after reclamation the original agricultural land use can be replaced by urban, industrial or horticultural use.

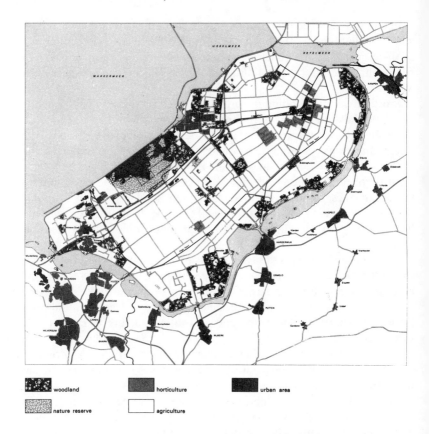

woodland

nature reserve

horticulture

agriculture

urban area

FIG. 2. Flevoland polder in the Netherlands.

3 SOILS

Most of the polder areas are found in regions with the following soil types: gleysols, fluvisols, histosols, or vertisols.[6] The permanent or seasonal wetness of these soils greatly influences their physical and chemical characteristics related commonly to their physiography. Under these conditions several factors of negative influence on the reclamation process and the agricultural productivity have to be taken into account:

(a) physical limitations: low-bearing capacity, causing settlement and instability of slopes forms serious constraints for reclamation, construction of embankments, canals and roads, foundations of structures and houses and agricultural land use (workability of the

land, accessibility of the land for cattle and machines); permeability, if high not suitable for irrigation, if low very hard to drain;
(b) texture of soils: swelling–shrinking, rootability;
(c) chemical properties: oxidation of peat, acid sulphate soils (cat clay), salinity, soil toxicity;
(d) conditions for germination;
(e) nutrient availability;
(f) seepage;
(g) oxygen availability.

During reclamation several measures may have to be taken, depending on the local conditions, to improve the texture of the soil and make it suitable for agriculture, such as lowering of the ground water table, leaching, soil improvement measures, application of chemicals.[7]

Especially in situations of permanently high water, the soil will be filled with water and have an unsuitable texture thus causing many problems for the first farmers trying to grow dry food crops. The ground water table must therefore be lowered.

A good example of possible measures are those taken in the new IJsselmeerpolders in the Netherlands to let the soil ripen.[8] They consist of land drainage together with an adapted crop rotation scheme (see Fig. 3). A first lowering of the ground water table is achieved with trenches, and crack formation in the clay soils starts resulting in an increase in permeability. When, after several years, the ground water table is deep enough (0.60 m below surface), the trenches are replaced by subsurface drains further lowering the ground water table. Since the bearing capacity of the soils during reclamation is very low, only crops like rape seed and cereals are grown, where cropping measures can take place during periods with an evapotranspiration surplus.

Leaching has to be applied when the soils are saline or alkaline, either naturally when the rainfall surplus during some period of the year is sufficient or artificially by surplus irrigation.

During reclamation, soil improvement methods such as deep ploughing, loosening, etc. may be required to establish good, structured, soils. Land levelling may also create optimal conditions for drainage and possibly also irrigation.

A special type of soil improvement for urban and industrial use, like local or integral landfill, may be required to create sufficient bearing capacity and drainage conditions. In most countries where urban or industrial development takes place in polders it is still cutstomary to raise

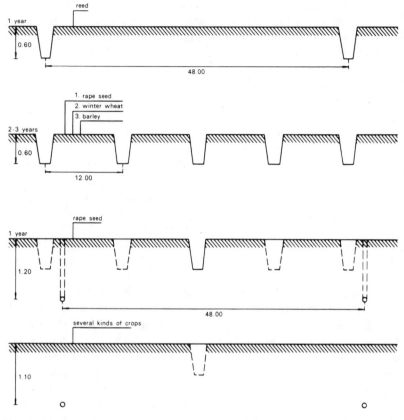

FIG. 3. Measures taken during reclamation in the IJsselmeerpolders in the Netherlands.

those lands above a certain level for safety reasons.

Application of chemicals may be necessary, depending on the soil condition and genesis during reclamation, e.g. the application of lime to acid sulphate soils.[9]

4 WATER MANAGEMENT

4.1 General

In a polder the primary function of the water management system is drainage, i.e. the temporary storage and eventual discharge of the water.

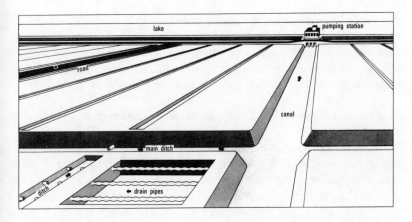

FIG. 4. Schematic layout of the drainage system in a polder.

The *drainage system* in a polder may consist of the following parts (see Fig. 4): *land drainage system*—subsurface drainage or open field drains, ditches; *hydraulic transport system*—ditches, main ditches, canals, culverts and weirs; *outlet structure*—sluices or pumping stations.

The design of the drainage system of a polder has to be based on the meteorological and soil conditions and the degree of control required.

The location and sizes of the different parts of the drainage system are not always exclusively determined by hydraulic considerations within the polder, but also by other factors such as agricultural economy, navigation, discharge of waste water, specific ecosystems, parcelling etc. Low and high tide, or river regime outside the polder, may influence the dimensions and the operation of the drainage system inside.

Some characteristic features of the drainage system that may vary according to local conditions are: capacity of the land drainage system; discharge capacity of the hydraulic transport system; polder water level; percentage of open water; pumping or sluice discharge capacity. These features are directly related to each other.[10] Thus a lower pumping capacity will result in a higher percentage of open water being required.

In a polder several types of *irrigation system* can be applied. When the surface level of the polder is relatively low compared with that of the adjacent water, irrigation by gravity is possible, which may at least partly be combined with the drainage system (mostly the hydraulic transport system but sometimes also the inlet and outlet structure are common). If salinity or alkalinity problems can be expected then the systems have to be separated. The irrigation system in a polder may be composed of the following parts: *field irrigation system*—basin, furrow, sprinkler, drip; *main irrigation system*—tertiary canal, secondary canal, primary canal; *inlet structure*—sluice, pumping station, well. The water resources for the irrigation system in a polder may be from surface or ground water.

4.2 Factors Influencing the Design of the Water Management System in a Polder

The factors influencing the composition and dimensions of the water management system in a polder are: meteorology, management of water outside the polder, land use, soil, reclamation, geohydrology, and water quality. These aspects of the design of the water management system in a polder will be discussed below with major attention being paid to the drainage system.

4.2.1 Meteorology

Of the meteorological conditions, rainfall in particular, evapo(transpi)-ration, temperature and wind are of importance, the first two by influencing the required dimensions of the drainage and/or irrigation system. Temperature is of importance for the growing of different crops, wind in the dike design. Of rainfall and evapotranspiration, the following are of importance: annual rainfall and evapo(transpi)ration and distribution over the year, short-term rainfall intensity, and the areal reduction effect of the rainfall.

The difference between rainfall and evapo(transpi)ration in the case of a surplus is decisive for the total amount of water to be discharged. In the case of a deficit, it determines the need to apply a drainage, as well as an irrigation system. Short-term rainfall extremes which can vary widely from place to place are important in the design of the dimensions of the drainage system. For the design of the drainage system of a polder, rainfall duration frequency curves are nowadays often used. To give an idea of their variation, Fig. 5 shows the curves for three widely differing locations.

The areal reduction effect of the rainfall is different in different climatological zones; local topography can also play an important role.[11]

4.2.2 Management of Outside Water

The conditions and management of the water outside the polder can be of importance in the design of the dike around the polder and the discharge structure. Regular fluctuations are especially of importance in the design

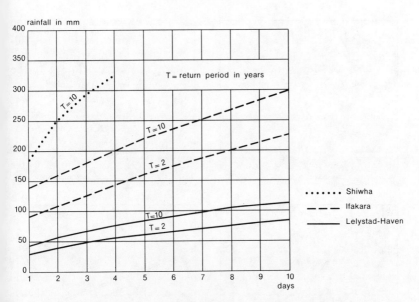

FIG. 5. Rainfall duration frequency curves for Ifakara (Tanzania), Lelystad-Haven (the Netherlands) and Shiwha (Republic of Korea).

of discharge sluices and the determination of the lifting device in the pumping station. Extreme situations are of importance in the design of dikes.

Flood protection and the construction of reservoirs upstream can also influence the design of dikes and discharge structures in the polder.

4.2.3 Land Use

The types of land use and their distribution over the polder can influence parts, or the whole, of the water management system.

Land use planning comprises the determination of types of land use, fixing sites and sizes of villages and divisions of the agricultural area into lots and farms. In areas to be reclaimed there is often no historically evolved situation with regard to human settlements or land title, which may therefore be parcelled out in the most economic way, choosing lot sizes best fitted to future farm sizes. There are several systems of parcelling out, easily to be distinguished in the landscape (see Fig. 6).

It is possible to choose the length–width relation for minimal costs per hectare for all cost factors depending on lot dimensions. In arable farming

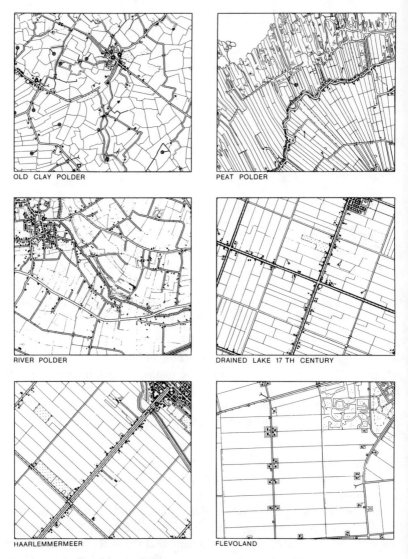

OLD CLAY POLDER

PEAT POLDER

RIVER POLDER

DRAINED LAKE 17 TH CENTURY

HAARLEMMERMEER

FLEVOLAND

FIG. 6. Different systems of parcelling out.

the following factors depend upon lot length and/or lot width:

— construction and maintenance costs of roads, ditches, main ditches, and canals;

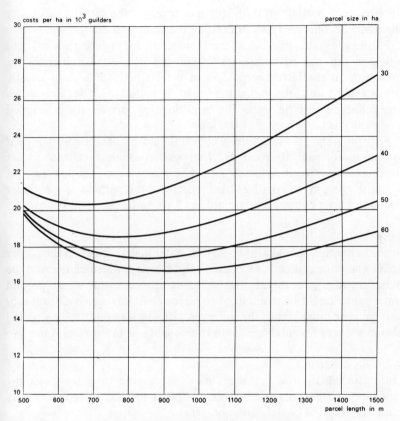

FIG. 7. Calculation of optimal plot sizes.

— yield depression on levelled spoil banks of ditches and main ditches;
— maintenance costs of slopes of main ditches;
— road ditches and parcel ditches;
— costs for installation and maintenance of subsurface drains;
— costs for the idle hours of tractors and machinery and commuting of workers;
— costs for transport of fertilizers, sowing seeds and harvest;
— yield depression on headlands and along parcel boundaries;
— turning costs of machinery.

The best lot length and lot width is a function of the lot size (see Fig. 7). For lots of 30 ha in this example the best dimensions are 450 × 700 m;

for 60 ha, 600 × 1000 m. With still greater length the costs increase only slightly. For different types of farming the principles are the same.

Farmers usually live in villages, probably for safety reasons, and although this has advantages for social life it is disadvantageous for transport. In the IJsselmeerpolders in the Netherlands a farm usually consists of 1.5 or 2 lots, so that the size of the standard lot has a considerable influence on farm sizes. The farm buildings are usually positioned along the road in a corner of the lot.

Regarding the infrastructure, a choice should be made between a road system and/or navigation canals. If navigation canals are chosen, either the drainage or irrigation canals should be made suitable for this purpose. In considering the alternatives, their costs and the purpose of the polder (small-holdings polder or estate polder) have to be taken into account.

4.2.4 Soil
Special attention should also be paid to water management, in the polder areas where the soil consists of acid sulphate clay, which can reduce the harmful effects of acidity. In the beginning of reclamation a high ground water table should be maintained to be lowered only slowly during subsequent years. Flushing of the soil, even with salt water, after a dry period also has a very favourable effect on the quality of the ground water.[7]

4.2.5 Reclamation
During reclamation several steps may be required to convert swampy lands into fertile soils which from the point of view of economy may influence the final layout of the polder.

Reclamation may be carried out in a relatively short period, over some years, or gradually as a long-term process, such as in the lowland development strategy in Indonesia. This strategy is based on several national development objectives, such as resettlement of people from the densely populated islands to unoccupied areas and increase of food production with emphasis on rice growing. The reclamation of tidal coastal lowlands is here characterized by a step-by-step development initiated by quick-yielding drainage projects based on low input and simple technology, to be followed by higher input and more sophisticated technology after the former stage has reached its optimal economic development, until the most modern and highest input technologies have been applied to achieve the maximum natural and socio-economic resource utilization. At this stage the level of advanced polders has been reached. The reasons for the step-by-step development are the lower initial costs needed to reclaim the large areas, and the time allowed. Another reason is that all the people concerned become familiar with the reclamation of swampy lowland. Features of the various stages with regard to water management infrastructure and kind of land use are as follows:

(1) *First stage*: open uncontrolled drainage system; rain-fed agriculture; one food crop annually; subsistence farming; main transportation by boats;

(2) *Intermediate stages*: semi-closed drainage system; semicontrolled water management based on flood and saline water protection; flushing in the case of acid sulphate soils and water conservation by surface and subsurface water control; mainly rain-fed agriculture; more than one crop annually; intermediate economic land farming; agro-processing industry and internal road system; trans-area transport;

(3) *Final stage*: closed system/fully controlled water management/ pumping; irrigation system; multiple cropping and integrated farming; commercial farming; sustained economic growth; industries and complete road-infrastructure system.

In this type of lowland development the water management system is based in the first place on drainage of tidal lowlands by gravity, making use of the tidal movement. Since, in the initial stage, only one rain-fed crop can be grown, farm sizes of 1·75–2·25 ha enable farmers to reach subsistence level (see Fig. 8). When the areas are developed further, the initial system will always be the starting point for further adaptations. Irrigation potential in particular can be reduced when too much attention is paid in the beginning to drainage only.

4.2.6 Geohydrology
The level of the surface water in a polder is generally low compared with the water table in adjacent areas, often resulting in seepage and damage to the crops when the seepage water is saline. Lowering of the ground water table in adjacent lands may also cause damage there (drying out of the land, subsidence, etc.).

4.2.7 Water Quality
Water quality plays a role both inside and outside the polder. Inside, salinity or alkalinity in particular can cause damage to the crops in the root zone. Outside, the required water quality in the receiving water body may put constraints on the discharge possibilities. If, for example, the

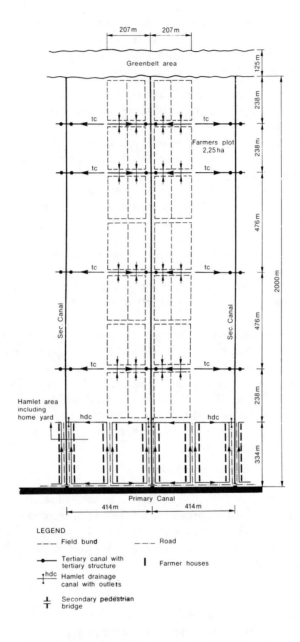

FIG. 8. Basic farm size, tidal lowlands, Indonesia.

water in the polder has a high salinity content due to saline seepage and the receiving water body has to be kept fresh, special measures have to be taken to prevent 'pollution of the receiving water body'.

4.3 Design of the Drainage System

When designing the water management system of the polder, two design approaches may be followed: empirical design and optimization. Empirical design is normally based on the conditions that have occurred in the past from which general data have been deduced. In the optimization approach the investments and maintenance costs of the water management system are compared with the yields and damage that can be expected in relation to the functioning of the system.[12]

The requirements of the land drainage system form the starting point in the design. For rice, the normal depth of flooding should not be exceeded too many times by a certain depth. For dry food crops the depth of the ground water table below the surface has to be controlled.

From the hydraulic characteristics of the land drainage system the amount of water to be evacuated can be derived to obtain the required degree of water control. Outlet capacity depends very much on the admissible storage in the polder.

Storage may be on the field, in the soil and in the open water system. Criteria for storage on the field, e.g. for rice crops, will largely depend on the type of crop and the stage of crop development. For old rice varieties a regular variation of the water layer on the field of 0·05–0·10 m is allowed. In the case of high-yielding varieties of rice the admissible surface storage on the fields is about 0·10 m with a return period of 5–20 years. Storage in the soil is important for dry food crops in temperate climates and may amount from 50 to 200 mm after a dry period, without raising the ground water level above a required depth. Especially significant is surface storage in the open canals, pools and lakes of the polder area. In areas under cultivation a permissible rise of the polder water level may vary from 0·10 to 0·40 m.

The design of drainage systems requires geological, topographic and soil maps, and climatic and agricultural data. Generally the investigations require a sequence of studies with increasing intensity. Therefore two or three phases in the investigations can be considered: (a) *reconnaissance level* (pre-feasibility study)—the main objective is to identify the feasibility of the proposed project, first of all on technical, but also on economic grounds; studies at this level are mainly based on existing information but may also include some field work; (b) *semi-detailed level* (feasibility study)

—alternative plans obtained from the reconnaissance study are worked out to a preliminary plan so that the competent authorities can make a decision; the data are the same as for the reconnaissance level, but are needed in more detail; (c) *detailed level* (project designs)—design of the selected project, including a list of quantities and preparation of tender documents.

Regarding the different data to be used in the studies, three types of variability have to be taken into account:[13] variability caused by inaccuracies of the method applied, variability in space, and variability in time.

4.3.1 Design of the Land Drainage System

Land drainage in polders is to prevent the occurrence of excessive moist conditions in the root zone. In arid areas, a further objective is to prevent the accumulation of salt in the root zone, or to leach accumulated salts out of the soil profile.

For the design of the land drainage system the reconnaissance level is used to investigate the need for drainage, as well as the type of drainage system. At the semi-detailed level an estimate is made of the costs, thus also the necessary drain spacings have to be estimated. In general the following information is used to make a land drainage plan:

— elevation, preferably contour intervals of 0·25 m or less;
— topography and infrastructure on maps 1:2500 or 1:10 000 to indicate land use, obstructions like open drains, canals and roads, outlet possibilities;
— soil texture to decide on drain depth and to estimate hydraulic conductivity;
— depth of impervious or impeding layers;
— hydraulic conductivity especially of soil layers between 1·0 and 2·5 m depth;
— ground water tables and fluctuations;
— ground water potentials at different depths;
— salinity of irrigation water and ground water;
— soil salinity and alkalinity;
— infiltration rate.

The ground water table is usually curved, its elevation being the highest midway between the drains. The factors influencing the height of the water table are:

— precipitation, irrigation and other sources of recharge;
— evapotranspiration and other sources of discharge;

— soil properties;
— depth, spacing and diameter of the drains;
— water level in the drains.

These factors are interrelated by drainage equations based on the assumptions of two-dimensional flow and of a uniform distribution of the recharge, steady or non-steady, over the area between the drains.

For the design of open field drains or subsurface drains, different steady state or non-steady state formulae are available such as those of Hooghoudt, Ernst, Glover-Dumm etc.[14] The steady state formulae are derived under the assumption that the recharge intensity equals the drain discharge and that the water table consequently remains in position. The non-steady state approach considers the fluctuations of the water table in time under the influence of a non-steady recharge. For steady state ground water conditions the drainage criterion has two parameters: height of the ground water table midway between the drains and the discharge. The other parameters in the formulae represent the soil characteristics and the spacing and diameter of the drains. For non-steady state ground water conditions the criteria are formulated in terms of a required rate of lowering of the ground water table. The appropriate choice of drainage criteria will depend on the following set of conditions:

— hydrological conditions determining the quantity of excess water to be drained within a specified time;
— agronomic conditions determining the permissible upper limit of the water table and its duration;
— soil conditions determining the relations between aeration and moisture content, ground water level and soil moisture content, ground water level and capillary rise;
— economic conditions determining the cost/benefit ratio between installation costs and benefits of better drainage conditions.

Since these conditions are complex and interrelated, the drainage criterion should be regarded as a simple approach to express the aims of a land drainage system in one or two simple parameters.

4.3.2 Layout of the System
Open field drains are normally installed parallel to each other perpendicular to the ditches. In subsurface drains three types of systems can be distinguished: random or irregular systems used to drain depressions or for interception of seepage in dikes; single parallel pipe drain systems,

where each subsurface drain has its own outlet to a ditch; composite pipe drain systems with subsurface drains and collector drains.

In homogeneous soils the greater the depth of the subsurface drains the wider their spacing. In practice the installation depth is limited by: installation costs, presence of low permeable layers, and the water level in the open drainage system. In humid areas the depth is usually between 0·90 and 1·20 m below the surface. In irrigated areas greater depths are chosen between 1·50 and 2·00 m below surface, especially for 'salt drainage'.

The slope of the drains in flat areas is generally 0·10 m per 100 m, and the maximum length of subsurface drains between 250 and 300 m (because of maintenance). For collectors slopes vary from 0·03 to 0·06 m per 100 m with lengths of up to 500 m.

The layout of the system is mainly dictated by the infrastructure and the position of the outlet facilities.

4.3.3 Discharge Capacity of Subsurface Drains

The diameters of subsurface drains and collectors must be such that their capacity is sufficient without considerable surcharge. For design, the most convenient way of computing the head-loss in drainage pipes is to use expressions relating the maximum drainable area A (ha), the drainage duty q (m/day), the average hydraulic gradient i, and pipe diameter d (m). For corrugated PVC pipes, Ven and Dekker[15] arrived at:

$$A = \frac{2270}{q} d^3 i^{0·666}$$

For smooth pipes, Wesseling[14] found:

$$A = \frac{769}{q} d^{2·714} i^{0·572}$$

The equations are derived for full flowing clean pipes and linear increase in discharge from 0 at the upstream end to maximum at the outlet. The available hydraulic gradient is usually taken as equal to the drain slope (although in reality the hydraulic head-loss line is curved).

For pipes with a constant flow throughout, the hydraulic gradient is a straight line. For a smooth transport collector the following applies:

$$A = 432 d^{2·714} i^{0·572}$$

The maximum drainable area in this case is smaller because the full flow has to be transported over the entire length. In order to compensate for sedimentation inside the drainage pipe it is common practice to reduce the maximum drainable area to 75% of the calculated area. A similar safety factor should be obtained when the available hydraulic head is lowered by 40% allowing pipes to be silted up to 25% of their diameter without consequences for the discharge capacity. For collectors it is common practice to increase the pipe diameter in the direction to the outlet, again reducing the maximum drainable area to 75% in practice.

4.3.4 Design of the Hydraulic Transport System

In truly level areas without any general slope, the design of drainage canals poses a real problem. The hydraulic gradient required for the flow can only be generated by a gradient in the water level artificially created by a drawdown at the outlet (sluice or pumping station) and by allowing the water level in the canal at its upper end to rise. The total of both heads is limited to about 0·3–0·6 m. This means that, if the length over which the water has to be transported through the canal is around 10 km, the available hydraulic gradient is only 0·03–0·06 m/km. Therefore the design velocities of drainage canals in level areas are necessarily small, of the order of 0·3–0·4 m/s. For the design of the hydraulic transport system the reconnaissance level is used to investigate the required principal layout. The information required for design of the hydraulic transport system is more or less the same as that for the land drainage system, with the addition of data on soil mechanical aspects and geohydrology of the area under study.

During the reconnaissance level and the semi-detailed level, calculations are normally based on hydrological routing models, such as the rational formula, the Muskingum method, reservoir models or steady state hydraulic formulae.[11] At the detailed level, hydraulic non-steady flow models are nowadays increasingly applied, especially when urban and industrial areas are involved.

4.3.5 Layout and Discharge Capacity of the Hydraulic Transport System

A complete hydraulic transport system consists of ditches, main ditches and canals. If a composite subsurface drainage system is installed, the collector drains replace the ditches. Normally the distances between ditches and main ditches are based on agricultural economy, resulting in optimal plot sizes. The canals are located so that a minimum of earth movement is required. The possible locations for sluices or pumping stations also determine the principal location of the canals.

The discharge capacity of the hydraulic transport system is normally such that a prescribed water level in the polder is not exceeded during a certain time at a certain return period. This, together with the accepted velocity in the different parts of the system, determines the cross-sections

4.3.6 Design of Sluices and Pumping Stations

Drainage by gravity rather than by pump lift is the most attractive solution for tidal parts of rivers, or tidal estuaries with a sufficient tidal range, or alongside rivers. A difficulty when applying this principle to polders along rivers lies in the fact that, in general, periods with heavy local rainfall coincide with periods with flood stages of the river, making gravity discharge temporarily impossible. The solution is to design a drainage canal parallel to the river with a gradient smaller than the slope of the river.

The discharge capacity of the sluices or pumping stations has to be calculated in combination with the discharge capacity of the canals, the available storage, and the accepted level fluctuations (see Table 2 for illustration).

TABLE 2

INDICATION OF THE REQUIRED DRAINAGE PUMPING CAPACITY FOR AGRICULTURAL AND URBAN AREAS IN RELATION TO THE OPEN WATER STORAGE IN THE REPUBLIC OF KOREA

Percentage of open water	Pumping capacity (mm/day)	Water level rise (m) at:	
		$T = 10$ years	$T = 100$ years[a]
Field storage is 100 mm (rice)			
12	22	1·04	2·03
16	12	1·03	1·82
20	10	0·88	1·54
Runoff coefficient of 0·8 (urban area)			
12	34	1·00	1·76
16	22	1·00	1·64
20	12	1·00	1·53

[a] These are theoretical values; inundation will occur in practice.

4.3.7 Main Drainage System Markerwaard Polder

An illustration of the determination of layout and discharge capacity of the hydraulic transport system may be found in recent studies for the system of the Markerwaard polder, the fifth polder of the Zuiderzee project in the Netherlands.[16] The following relevant information was available:

(1) The main type of land use will be agriculture (> 50% of the total area). Depending on the soil suitability, arable farming, dairy farming, horticulture and fruit growing are planned, and considerable parts to be used for afforestation and the creation of nature reserves.

(2) There is a large variation in the texture of the top soil, with medium fine to locally coarse sand in the north-eastern part gradually changing into loam and clay in the south-western direction. Locally old marine clay and peat are present.

(2) Based on calculations with computer models for ground water flow (finite elements method), average seepage after reclamation is estimated at 0·7 mm/day.

(4) In the northern part, the deeper ground water has a chloride content ranging from 1000 to 3000 mg Cl/litre; on the south-western part it is 100–1000 mg Cl/litre.

(5) The elevation of the lake bottom at emergence ranges from about 2·0 to 4·8 m below mean sea level. Because of ripening and dewatering of the loosely packed marine sediments, considerable subsidence will occur after emergence, accurate prognosis being possible on the basis of the available data. For the design of the drainage system, surface levels 80–100 years after emergence are taken as decisive.

Boundary conditions for the design. Based on information and practice in the existing IJsselmeerpolders, the following items were taken into consideration:

(a) For agriculture the minimum required depth of the subsurface pipe drains at the outlet is 1·10 m below the soil surface, and for fruit orchards 1·30 m. To avoid frequent submerging of the subsurface drain outlets, the polder water level should be at least 1·40 m below the soil surface. A freeboard of 0·30 m is sufficient to account for flow resistance and temporarily higher water levels due to wind effects and rainy periods.

(b) The acceptable exceedance frequency for the polder water level in the main drainage system is: 0·40 m once in 2 years; 0·65 m once in 10 years.

(c) The pumping station(s) should be located at the south-eastern dike of the polder because of the salt load. From here the water pumped out can be discharged via Amsterdam to the North Sea, thus minimizing negative effects on the quality of the fresh water in the surrounding lakes.

(d) The drainage system may be designed for agriculture, with respect to the polder water level and pumping capacity, thus giving enough flexibility for final land use planning.

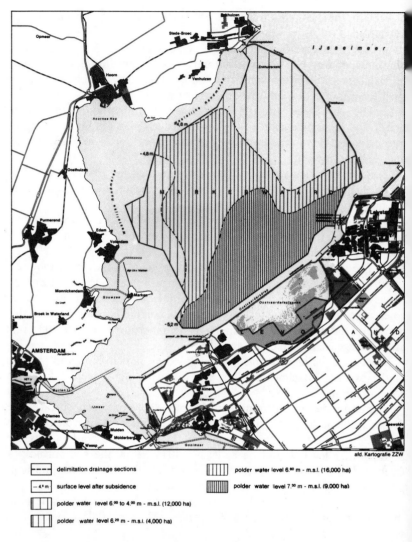

FIG. 9. Main polder sections in the Markerwaard.

The polder water levels. With respect to level, the polder comprises four parts: elevated areas in east and west, a very low southernmost part and north of that a large central strip of intermediate height. It may be concluded that about 25% of the Markerwaard located in the central strip will have an elevation between 5·0 and 5·2 m below mean sea level. Therefore the main polder section should have a water level of 6·6 m below mean sea level (Fig. 9). The deeper southernmost part should have a polder water level of 7·3 m below mean sea level. The higher elevated part in the west may get a level of 6·2 m below mean sea level to reduce the excavation costs of the open drains. In the higher elevated north-eastern part of the polder, with light loamy to sandy soils, different polder water levels are recommended. Depending on the final land use, a smaller or larger intake of water will be necessary, especially for agriculture. In this area the water levels during the growing season will be about 0·40 m higher than in winter time. This higher elevated area can discharge the excess water by gravity to the main polder section at 6·6 m below mean sea level.

For the study of the required pumping capacity, data for the North-East Polder have been used, its size and most of the physical conditions being quite similar to those of the Markerwaard. For the North-East Polder, time series of daily amounts of water pumped out by the pumping stations (outflow) are available. To find the daily inflow of excess water into the hydraulic transport system, the daily changes in open water storage are added to outflow data by using water level recordings at different points in the canals and the known relation between water level and storage in open canals. The inflow data are corrected for seepage and intake of water from outside the polder. With these data a frequency analysis with Gumbel's extreme value method has been applied. The data have been used to find the required pumping capacity for the Markerwaard by using the following water balance equation:

$$I + S = Q + B$$

where I is the net inflow of excess water into open drains (mm), S is the seepage inflow (mm), Q is the discharge of the pumping station (mm), and B is the storage capacity in the open drains (mm).

The calculations were made for different durations and frequencies once in 2 and 10 years. A percentage open water of 1% was considered to be a minimum in order to realize the required discharge capacity and navigation requirements. A higher percentage was not economic because of land losses. The calculated net pumping capacity is 11·9 mm/day or 57 m³/s. The frequency of once in 10 years is decisive. The calculated capacity is based on 100% reliability and direct switching on of the pumps in wet periods.

For the layout of the hydraulic transport system, four alternatives have been studied (Fig. 10) and the alternatives compared with respect to water

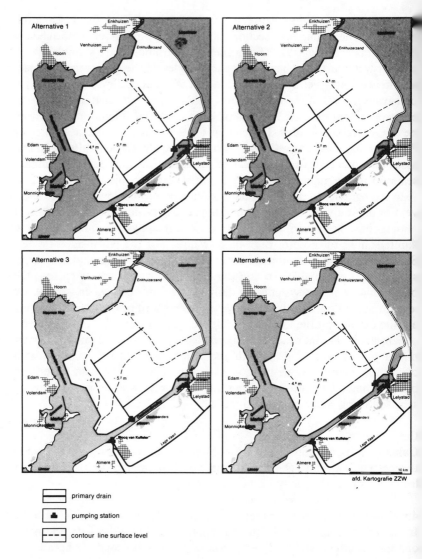

primary drain

pumping station

contour line surface level

FIG. 10. Alternative layouts for the hydraulic transport system of the Markerwaard.

management aspects, investment and maintenance costs, vulnerability (demolition, sabotage), navigation and location of shiplocks, and distribution of brackish seepage in the polder. Taking all these factors into consideration, preference was given to one pumping station (alternative 4 on Fig. 10). The pumping station will be equipped with a common type of pump units with a capacity of 11·5 m³/s each. To realize the net pumping capacity of 57 m³/s, five units would suffice, but to have a safety margin and to account for maintenance, six units with a total capacity of 69 m³/s or 14 mm/day will be installed. This capacity is in fair agreement with an evaluation of the water management in the existing polders. Two units will be used for the lowest polder section (9000 ha) with a polder water level 7·3 m below mean sea level. One of these two units can also be used to assist the other four units to maintain the water level at 6·6 m below mean sea level. As energy sources both electricity and diesel will be used.

Dimensions of primary drains are determined by the maximum flow velocity of 0·25 m/s at the design discharge.

4.4 Irrigation and Drainage

Since polders are provided with a detailed system of drainage canals, it should be examined whether the same system can be used for the supply of irrigation water. If this is the case the water will have to be lifted to field level but, since in many large level areas the irrigation water has to be lifted anyway, this is not a real drawback.

When irrigation water contains some salt and evapotranspiration dominates, no salt is removed from the system and an accumulation of salt occurs in the root zone (primary salination). If the ground water is brackish and at shallow depth (less than around 2 m), capillary ascent supplies still more salt to the root zone.

To control salinity an additional amount of irrigation water in excess of the consumptive use is applied (10–20%) to leach the soil by percolation. Capillary ascent is counteracted by keeping the phreatic table well below the surface (1.5–1.8 m) by means of deep subsurface drainage. The leachate has to be removed by the drainage system.

4.5 Flood Control and Flood Protection

In many cases protection from flooding is a first step in the reclamation of waterlogged lands either from the sea or from rivers. In the case of rivers, protection may be achieved by flood control in the upstream part of the river, with flood retention reservoirs as the most effective means. Flood levels can also be reduced by measures near the lands to be protected, such as channel improvement (deepening, cut-offs, diversions) and by-passes.

Flood protection around polders means the erection of dikes, which may cause side effects. Side effects of a hydraulic nature occur when before embanking, the river overtops the banks. Elimination of the overland flow results in a rise of the flood levels. The effect is mostl pronounced in the case of flash floods with rapid rises. By aligning the dikes away from the channel, a larger cross-sectional area can be obtained and some storage is recovered but the strip of the natural levee between the channel and the dike is now exposed to more flooding than before.

Dikes may entail morphological effects by preventing overbank spilling and the deposition of sediments which contribute to the building-up of the flood plains and are also beneficial as soil dressing. As a result, the rive has to carry a greater amount of sediments than before. If the river, in spite of an increase in the peak flow, is unable to do so, the river bed, will sil up, leading to a further rise of flood levels.

Embanking of rivers and subsequent change of the river regime promote meandering so that the dikes are threatened by erosion. Since river training is very expensive, the dikes have to be aligned at a sufficien distance from the meandering channels to avoid an early failure.

There are several side effects related to the environment and the water control of the land areas. Floods have a beneficial effect in that they periodically remove accumulated human disposal and dirt, and saline water is also washed away. After embanking, the water is stagnant for most of the time and flushing of the water courses with river water may be necessary.

4.5.1 Principles of Dike Design

Dikes may be divided into sea and river dikes and dikes along canals or inland waters. In the past the crest height of the dike was determined by the highest known flood level. Little was known about the relationship between the cost of preventing flooding and the cost of the damage that might result from flooding. Modern design methods utilize knowledge obtained about forces acting on a dike and the strength of its structural elements and probabilistic design methods.[17] The inundation risk of the polder or the probability of failure of a dike section are especially of importance. Studies are nowadays generally carried out in two directions:

— an assessment of the 'profit' related to the level of security involving an evaluation of the possible damage caused by the failure of a dike (both in financial terms and in loss of human life and valuable cultural sites and amenities). The aim is to fix standards of safety for specific polders in relation to particular defence requirements;

— studies for a well balanced design for a specific safety standard or risk of failure so that all elements of the dike bear the same risk.

The first step in dike design is an analysis of all possible causes of dike failure, comprising four categories of events likely to cause inundation: human failure and management faults, aggressive human actions such as war or sabotage, extreme natural conditions, and technical failure of structural elements.

The second step is to analyse the different causes of failure in a particular category. Only technical failures are mentioned here: overflow of the dike, erosion of the wetted slope or loss of stability, erosion of the inner slope leading to progressive failure, instability of the whole dike, and instability of the foundations and internal erosion. For all these modes of failure, the situation is considered where the forces acting are just balanced by the strength of the construction. The probability of occurrence may for each failure mechanism be found by mathematical and statistical techniques. Improved knowledge of the theoretical relation between wave attack and the strength of the revetment, the probability of slope stability related to the various ground parameters and the theory of internal erosion is required.

4.5.2 Sea Dikes

Several elements play a role in determining the design level of a sea dike:

— wave run-up depending on wave height and period, angle of approach and roughness of the slope;
— an extra margin to the dike height to take into account seiches and gust bumps (single waves resulting from a sudden violent rush of wind):
— a change in chart datum (MSL) or a rise in the mean sea level;
— subsidence of the subsoil and the dike during its lifetime.

Based on these factors the design level of a sea dike can be determined as illustrated in Table 3.

Nowadays the occurrence of extremely high water levels at sea can be described adequately in terms of frequency in accordance with the laws of probability. However, the curves of extreme water levels, based on a relatively short period of observations, have to be extrapolated into regions far beyond the field of observation. In several cases no practical level can be given which will never be exceeded, therefore there will always be some risk. In the Netherlands the design of all sea dikes is based on a water level with a probability of exceedance of 10^{-4} per annum.

TABLE 3
DETERMINATION OF THE DESIGN LEVEL OF SEA DIKES[17]

Flood level	MSL +	5·00 m
Wave run-up		9·90 m
Seiches and gust-bump		0·35 m
Rising mean sea level (MSL)		0·25 m
Settlement		0·25 m
Design level	MSL +	15.75 m

4.5.3 River Dikes

River dikes are generally made to withstand the highest flood levels. These dikes are often small, positioned along the natural levees and sometimes carrying roads. The present form and height of most of the river dikes is achieved by raising the crest only, using all kinds of locally available soil, and as a consequence the inner slope steepens increasingly. As a result the inner slope could become unstable during periods of high river water level.

In the past the construction of these dikes was more or less based on experience only, resulting in many disasters caused by inundation and failure. This made a better protection against inundation essential. Today, the incidence of extreme high water levels can be represented in terms of frequency, using the same philosophy as described for sea dikes. In the Netherlands it was decided to base the design level of the main river dikes on a water level corresponding to a discharge that can be reached or exceeded 8×10^{-4} times per year.

4.5.4 Dikes Along Canals or Inland Waters

In the Netherlands, 'belt' canals were excavated around the lakes to be drained. With the soil dug from these canals (often peat) a belt canal dike was made between the canal and the lake. These belt canals left part of the existing system of waterways intact, to transport superfluous water to the sea or river and for use by shipping. During the centuries of impoldering in the western part of the Netherlands an interconnected network of belt canals and thousands of km of belt canal dikes have been established.

The subsoil in this area consists mainly of strata of peat and soft clay, causing severe subsidence of the dikes at a rate of up to 0·05–0·10 m per annum in some areas. As the water level of the belt canal and the whole system of canals to the sea is kept constant, this has necessitated the frequent raising of the belt canal dikes (up to once every 2–3 years) usually using locally available materials and resulting in an inhomogeneous top layer of up to 4–5 m thickness of dredged mud, peat, clay, rubble, ashes and sometimes sand. Often the additional weight has produced further subsidence.

Safe belt canal dikes have become important since a growing population and industrialization have necessitated building in low-level polders. When a belt canal dike bursts, the low-level polder will be inundated completely by the water stored in the belt canal system, and in addition damage may occur which is difficult to repair. The dike burst causes a depression of the level of the adjacent belt canal, endangering the stability of the dikes alongside the canal.

The design of belt canal dikes has not changed very much over the years, as the major work on this sort of dike is maintenance. The crest must have a minimum width of 1·5 m, but 3 m is recommended to cater for vehicular transport. The crest must be 0·5–0·8 m above the extreme belt canal level which is regulated by the pumping station of the polders and the pumping stations and sluices that drain the belt canal into the sea, and which varies within certain limits. Nevertheless the design for a new cross-section must be based on geotechnical investigations and calculations and will often show flatter inner slopes.

4.6 Influences of the Polder on the Existing Hydrological Regime

In general the making of a polder can influence the existing hydrological regime in several ways, e.g.:

— lowering of the ground water table in adjacent lands;
— reduction of the cross-section of a river floodplain;
— increase of the salt water intrusion into the river in case of impoldering lowlands near the sea coast;
— after making dikes alongside the river, sediments will no longer be deposited in the river and coastal swamps. Apart from the lack of natural heightening, a subsidence process will also start in the adjacent polders, due to a lower drainage base compared with the original situation. In addition, the continuous process of river bottom elevation due to sedimentation will continue.

The last processes will lead to an increasing difference in river water level and polder water level, reducing the possibilities of natural drainage by sluices and increasing the seepage flow into the polder.

5 POLDERS OF THE FUTURE

To obtain an impression of the total area potentially suitable for impoldering, Van Diepen[18] has selected soil units from the 1:5 million FAO–
UNESCO soil map of the world,[6] which are related to conditions of
impeded drainage caused by flooding by river, or sea, or by ponding of
rainwater, or a combination of both. A second selection criterion used is
the topography of the area, choosing land with a slope of 2% or less. From
the resulting area were excluded:

— three gleyic soil units occurring in temperate zones and with water-
  logging problems restricted to the winter;
— soil units with indurated layers, hard rock at shallow depth, and with
  many stones or gravel at the surface;
— all soils of the permafrost zone.

The area where the potential polders have to be looked for has now been
reduced to approximately 1230 million ha. Further reduction can be made
when assuming that potential polders will not be constructed in areas with
problem soils, e.g. potentially acid sulphate soils, or Dystric Histosols
which will become very acid, with a pH less than 3–4 after drainage. But
more constraints can be considered relating to reclamation activities, such
as construction of dikes, structures and buildings or with respect to the
kind of land use after impoldering.

If the soil is confined to agricultural use after reclamation, three major
kinds of land use can be distinguished: dry food croppings, grassland, and
wetland rice. The suitability of the selected soil units has been evaluated
for the three kinds of agricultural land use based on the occurrence of
soil-related constraints. From this evaluation it can be learned that the
soil-related constraints of the Gleyic Cambisols, the Gleysols, the
Fluvisols and Luvisols are negligible to moderate for all kinds of agricultural land use, while the Podzols and the Solonchaks are moderately
suitable for wetland rice cropping. Taking into account all the unreliabili-
ties and uncertainties attached to the inventory, it may still be stated that,
from a soil suitability point of view, polders can be expected in an area
with a total area between 500 and 600 million hectares.

With regard to the socio-economic aspects related to polder development in the future, we can distinguish three different types of potential
polders:

— polder development in already densely populated areas, mostly with

small farmers, working under bad conditions of water management, soils and infrastructure;
— polder development in the sparsely populated level and wet areas, situated in a densely populated region;
— polder development in sparsely populated level wet areas, in a sparsely populated region.

In the first case the need for land is often such that the farmers themselves will try to reclaim it. While normally the best land is already occupied, the conditions will be difficult and government aid will often be required to improve the situation. The second type normally requires a large scale approach, initiated by central government. The government reclaims and develops the land and rents or sells it to farmers. The third type normally also requires a governmental approach. People have to come from other parts of the country and, having no relation with the new area, a complete new society has to be established. In general, big social problems face these projects in the initial stage.

With regard to the environmental aspects, land reclamation projects greatly influence the existing environmental regime. Nowadays, in the framework of studies to be executed for new projects, an environmental impact analysis will also be carried out, generally consisting of two parts:

— the influence of the project on the existing environment, and measures to be taken to reduce this influence as much as possible;
— new environmental values to be developed within the framework of the project.

REFERENCES

1. ZONN, I. S. and NOSENKO, P. P. Modern level of and prospects for improvement of land reclamation in the world, *ICID Bull.,* **31**(2) (July 1982).
2. VOLKER, A. Lessons from the history of impoldering in the world. In: *Polders of the World* (final report), International Institute for Land Reclamation and Improvement, Wageningen, 1983.
3. INTERNATIONAL COMMISSION ON IRRIGATION AND DRAINAGE. *Multilingual Technical Dictionary on Irrigation and Drainage,* New Delhi, 1967.
4. SEGEREN, W. A. Keynote: introduction. In: *Polders of the World* (final report), International Institute for Land Reclamation and Improvement, Wageningen, 1983.
5. SOKOLOV, A. A., RANTZ, S. E. and ROCHE, M. *Flood Flow Computation* (methods compiled from world experience), Unesco Press, Paris, 1976.
6. FAO–UNESCO. *Soil Map of the World,* Rome, 1971–1979.

7. ILACO. *Agricultural Compendium for Rural Development in the Tropics and Subtropics*, Elsevier, Amsterdam, Oxford, New York, 1981.
8. RIJNIERSCE, K. A simulation model for physical soil ripening in the IJsselmeerpolders, Flevobericht no. 203, IJsselmeerpolders Development Authority, Lelystad, 1983.
9. DOST, H. and BREEMEN, N. *Proc. Bangkok Symp. Acid Sulphate Soils,* International Institute for Land Reclamation and Improvement, Wageningen, 1982.
10. SCHULTZ, E. A model to determine optimal sizes for the drainage system in a polder, *Proc. Int. Symp. Polders of the World,* Vol. I, International Institute for Land Reclamation and Improvement, Wageningen, 1982.
11. CHOW, VEN TE. *Handbook of Applied Hydrology* (a compendium of water resources technology), McGraw-Hill, New York, 1964.
12. INTERNATIONAL COMMISSION ON IRRIGATION AND DRAINAGE. *The Application of Systems Analysis to Problems of Irrigation, Drainage and Flood Control* (prepared by the permanent committee of the ICID on the applications of drainage and flood control), Pergamon Press, Oxford, 1980.
13. WORLD METEOROLOGICAL ORGANIZATION. *Guide to Hydrological Practices,* Vol. I, Data acquisition and processing, 4th edn, WMO no. 168, Geneva, 1981.
14. INTERNATIONAL INSTITUTE FOR LAND RECLAMATION AND IMPROVEMENT. *Drainage Principles and Applications* (I, Introductory subjects; II, Theories of field drainage and watershed runoff; III, Surveys and investigations; IV, Design and management of drainage systems), Wageningen, 1974.
15. VEN, G. A. and DEKKER, K. The discharge capacity of subsurface drain pipes and its influence on the design of the subsurface drainage system (in Dutch), *Cultuurtechnisch Tijdschrift,* **22**(1) (June/July 1982).
16. SCHULTZ, E., SEVERS, G. J. and VEN, G. A. Design study of main drainage system Markerwaard, *Proc. Symp. Land Drainage,* Southampton, 1986.
17. MAZURE, P. C. The development of the Dutch polder dikes. In: *Polders of the World* (final report), International Institute for Land Reclamation and Improvement, Wageningen, 1983.
18. VAN DIEPEN, C. A. *The Flat Wetlands of the World* (a soil exploration to potential polder areas), International Soil Museum, Wageningen, 1982.

INDEX